NELSON

FIT FOR LIFE

WORKBOOK

HEALTH AND PHYSICAL EDUCATION FOR THE AUSTRALIAN CURRICULUM

LEVELS 9 + 10

2ND ED

ROB MALPELI
AMANDA TELFORD
CLAIRE STONEHOUSE
LEE ANTON-HEM
MICHAEL SPITTLE
SAM WATKINS
EMMÉ WILD
DAVID BAKKER

Nelson Fit For Life Health and Physical Education for The Australian Curriculum Levels 9 and 10 Workbook
2nd Edition
Rob Malpeli
Amanda Telford
Claire Stonehouse
Lee Anton-Hem
Michael Spittle
Sam Watkins
Emmé Wild
David Bakker

Product manager: Sarah Craig/Cathy Beswick-Davison
Content developer: Rachael Pictor
Project editor: Sutha Surenddar
Editor: Anne Mulvaney/MPS Limited
Proofreader: MPS Limited
Production controller: Karen Young/Sutha Surenddar
Permissions/Photo researcher: Liz McShane
Project designer : Mariana Maccarini
Cover designer: Mariana Maccarini
Text designer: Mariana Maccarini
Typeset by: MPS Limited
Cover: Shutterstock.com/Nicetoseeya
MPS Limited

Acknowledgements
We respectfully refer to Aboriginal and Torres Strait Islander Peoples as First Nations Peoples throughout the book.

ACKNOWLEDGEMENT OF COUNTRY
Nelson acknowledges the Traditional Owners and Custodians of the lands of all First Nations Peoples of Australia. We pay respect to their Elders past and present.
We recognise the continuing connection of First Nations Peoples to the land, air and waters, and thank them for protecting these lands, waters and ecosystems since time immemorial.

WARNING:
First Nations Peoples are advised that this book and associated learning materials may contain images, videos or voices of deceased persons.

For product information and technology assistance,
in Australia call 1300 790 853;
in New Zealand call 0800 449 725

For permission to use material from this text or product, please email aust.permissions@cengage.com

National Library of Australia Cataloguing-in-Publication Data
A catalogue record for this book is available from the National Library of Australia.
9780170465540

Cengage Learning Australia
Level 5, 80 Dorcas Street
Southbank VIC 3006 Australia

Cengage Learning New Zealand
Unit 4B Rosedale Office Park
331 Rosedale Road, Albany, North Shore 0632, NZ

For learning solutions, visit cengage.com.au

Printed in China by 1010 Printing International Limited.
3 4 5 6 7 26 25

CONTENTS

9780170465540

MAKING INFORMED DECISIONS ABOUT ALCOHOL AND OTHER DRUGS

Shutterstock.com/Nicetoseeya

WORKSHEET 1.1 WHAT DO YOU ALREADY KNOW?

Pages 4–5

How much do you already know about alcohol and other drugs? Complete the 'brainstorm' below with all your thoughts and ideas on this topic. Add 'bubbles' to the brainstorm. Start by identifying as many illegal drugs as you can and add these in bubbles around the central bubble; then add more bubbles for information you know about the drugs.

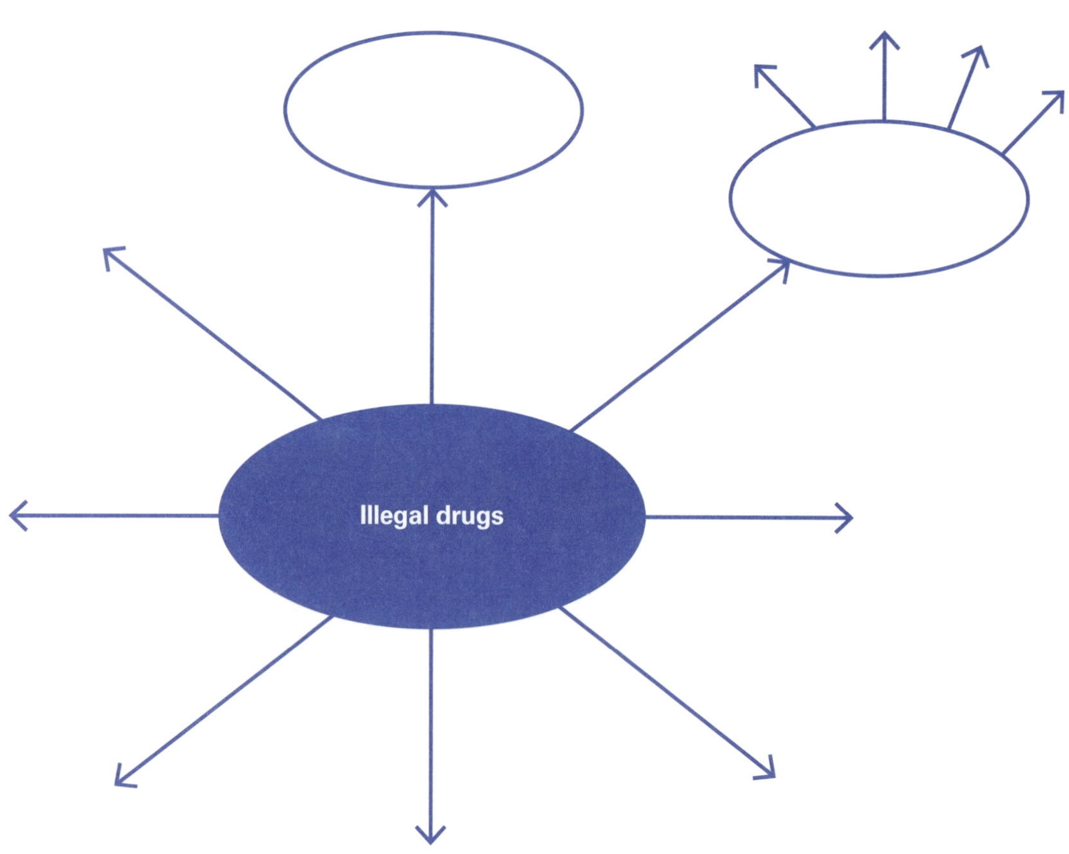

9780170465540

WORKSHEET 1.2 OPIOID CASE STUDY

Pages 6–7

CASE STUDY

A third of Australians surveyed in a global drug study said they used prescription opioids in the past year, the highest number of all countries surveyed. Globally, half of people using prescription opioids said they took them to get high. About three Australians a day die from drug overdoses involving opioids. It's a grim statistic that far outstrips the number of lives taken by heroin. Most opioids are given to people over 65. But the new survey data suggests many younger people are using them as well.

The data comes from 123 814 people who responded to the Global Drug Survey, released on Thursday. The survey was completed online and anyone could submit answers. About 8000 Australians responded, versus 35 030 Germans. Because of this, it cannot be used to estimate drug use in a population. The average Australian respondent was 23, male, white, and went clubbing four or more times a year.

Opioid painkillers – which come from a class of drugs including heroin – are often prescribed by doctors after an injury. But, like heroin, they are extremely addictive. More than three million Australians obtained an opioid prescription in 2016–17, according to data from the Australian Institute of Health and Welfare. About 715 000 people were using painkillers or opioids for non-medical reasons. In America, where opioids have been declared a public emergency, more than 130 people die every day.

'The idea that Australians would be using more prescription opioids than anywhere else, we should note that, but be cautious about what it means,' says Professor Michael Farrell, head of the National Drug and Alcohol Research Centre.

'People who choose to do the survey may have different behaviour than the whole population, so you need to interpret it with caution.'

Globally, the most-used drug – excluding alcohol – was cannabis (86 per cent of respondents) followed by MDMA (63 per cent) and cocaine (61 per cent). About 13 per cent of Australians said they had used some form of pill-testing. Australians reported getting drunk 47 times a year on average. And 38 per cent of people globally who reported drinking alcohol said they wanted to drink less. However, Professor Farrell urged caution in interpreting these results.

'One of the paradoxes in some of these findings is we know internationally and in Australia that young people are drinking less, getting drunk less, and waiting to be older to drink,' he said. Globally, there was a slight rise in the number of people who said they used cocaine and then needed emergency medical treatment versus last year's survey. This suggests a global increase in the purity of cocaine, which causes more harm, the study's authors said. The global average price per gram also rose from $107 in 2016 to $129 last year. Australians rated cocaine as the worst-value drug. Colombians, by comparison, think it is excellent value.

'Young Aussies taking opioids at highest rate in world, survey finds' by Liam Mannix, *The Sydney Morning Herald*, 16 May 2019.

WORKSHEET 1.2 CONTINUED

1 A third of Australians who responded to the Global Drug Survey said they had used opioids in the past year. Globally, half of those surveyed who used prescription opioids said they used them to get high. According to the article, approximately how many Australians die from opioid overdose each day? Calculate how many people would die each year from opioid overdose.

____

____

2 According to the article, identify the top three most used drugs globally (not including alcohol)? Rank the drugs in order and include the percentage rate of survey respondents.

____

____

____

explain
to provide extra information that demonstrates understanding of reasoning and/or application

AC

3 **Explain** why you believe Professor Michael Farrell, head of the National Drug and Alcohol Research Centre, urged caution in interpreting the results of the Global Drug Survey.

____

____

____

____

____

discuss
to talk or write about a topic, taking into account different issues or ideas

AC

4 Professor Michael Farrell mentioned that 'in Australia young people are drinking less, getting drunk less, and waiting to be older to drink'. **Discuss** reasons why you think young people are taking a more responsible approach to alcohol consumption.

____

____

____

____

____

____

5 Endorphins are chemicals produced by the body to relieve stress and pain, and make us feel good. Activities such as exercise, laughing and eating comfort food help the body to release endorphins. Opioids work similarly to endorphins by relieving pain and producing a feeling of euphoria.

Using the internet, investigate how opioids mimic the effect of endorphins. **Determine** why opioid abuse is so dangerous.

determine to establish, conclude or ascertain after consideration, observation, investigation or calculation; decide or come to a resolution

WORKSHEET 1.3 HOW DRUGS AFFECT THE BODY

Page 11

In order to understand why and how people use drugs, it is important to learn about the five main categories of drug use. There are five main types of illicit drug use. They are:

- experimental
- recreational
- situational
- intensive
- dependent.

Match the definition to the correct type of drug use.

Experimental use	A person's need to use a drug to make them feel normal
Recreational use	Short-term or 'one off' use of drugs to satisfy curiosity
Situational use	Excessive use of drugs over a short period of time
Intensive use	Using drugs to cope with the demands of a situation
Dependent use	Social or casual use of drugs, perhaps on a weekend or as part of a social life

Alamy Stock Photo/Phanie

WORKSHEET 1.4 IT'S OK NOT TO TAKE DRUGS

SB
Pages 30–1

Not everyone takes drugs. While there are many reasons people may consider taking drugs, there are also many reasons why people either choose not to take drugs or even stop their drug use. Personal, social, environmental/cultural and economic factors play an important part in the choices young people make about drug use. The influence of these factors can either be positive or negative. Any influence that affects your choice can either be a risk factor or a protective factor.

1 **Investigate** some of the reasons why a young person may choose to either start or not start using drugs.

investigate to carry out an examination or formal inquiry in order to establish or obtain facts and reach new conclusions; search, inquire into, interpret and draw conclusions about data and information

Risk factors Reasons to start taking drugs	Protective factors Reasons to not start taking drugs

2 **Decide** which protective factor you think would convince young people the most to *not* start using illegal drugs.

____

____

____

decide to reach a resolution as a result of consideration; make a choice from a number of alternatives

3 **Propose** alternative activities that a young person can consider instead of using drugs.

____

____

____

propose to put forward (e.g. a point of view, idea, argument, suggestion) for consideration or action

WORKSHEET 1.5 ALCOHOL MYTHS AND FACTS

Page 18

A myth is a false belief that is often assumed by many to be true, but in fact is not. A fact is based on hard evidence and has been proved to be true. Read the following statements about alcohol and determine whether you think the statement is a myth or a fact. Circle the correct answer and state the reasons to support your decision.

1 When partying, everyone gets drunk.
Myth/fact

____

____

2 Calling an ambulance means the police will be contacted.
Myth/fact

____

____

3 Drinking is cool and if you don't drink, you will be seen as the odd one out!
Myth/fact

____

____

4 Spiking a friend's drink with alcohol is illegal.
Myth/fact

____

____

5 Eating food will help to 'soak up' alcohol.
Myth/fact

____

____

6 If you don't eat, you will get drunk more quickly.
Myth/fact

____

____

9780170465540

WORKSHEET 1.6 EFFECTS OF DRUG USE

SB Page 26

Personal, social, environmental/cultural and economic factors all play a part in the choices young people make about using drugs. Their influences can either be positive or negative.

Your friend has started to experiment with LSD. You are concerned and want to help. Write a letter to your friend, expressing your concerns. Include diagrams if you think they might help. Include the following information:

- Discuss the personal, social, economic and environmental/cultural effects of taking illegal drugs.
- Describe the effects of LSD on the body.
- Explore alternative, healthier options.
- Recommend helplines or other ways your friend can get professional advice.

Dear,

WORKSHEET 1.7 THINK IT THROUGH

SB Page 29

Everyone enjoys a fun party or gathering where they can unwind and catch up with friends and family. In small groups, discuss and brainstorm responses to the following questions. You may use your student book to help you.

Describe your idea of having a good time.

You and your friends are invited to a party advertised on Facebook. You are all looking forward to going but you have some concerns. Identify and discuss the possible concerns that you may have.

Explain what you could do to ensure you and your friends have a great time and remain safe.

9780170465540

WORKSHEET 1.8 DRUGS AND DECISION-MAKING

SB
Page 32

The brain is the control centre of the body. Weighing around 1.3 kg, it controls everything that the body does. Drugs contain chemicals that tap into the brain's communication system, affecting a person's ability to process information and make sound decisions. Drugs can also change the way people behave and slow down the reflexes, which can lead to accidents and injuries.

Using your student book or research, answer the following questions.

1 **Explain** how drugs can affect your ability to drive.

explain
to provide extra information that demonstrates understanding of reasoning and/or application

2 Explain how drugs can affect your ability to make decisions.

3 Choose one of the following drugs and explain how the drug can increase the likelihood of risk-taking and injury:

LSD Alcohol Ecstasy Methamphetamine

WORKSHEET 1.9 WEIGHING UP THE OPTIONS

Page 33

THINK, PAIR, SHARE

In pairs, consider the effects of drug use. Read the following scenarios and decide what the best decision would be. When responding to each scenario, consider the following questions:

1 Identify the risks.

2 Discuss how might you be feeling in this situation.

3 Determine your options.

4 Decide on your decision and discuss reasons why.

5 If you had consumed alcohol, do you think your decision would have been different?

SCENARIO 1

You are a 17-year-old boy. You are at a house party and it's late. You had originally planned to go home with your mate, who has just got his licence, but he has been drinking. He insists that he is fine and he'll drop you off home. You know that a condition of his probationary licence is a zero blood alcohol concentration, but you don't want to create a scene. What do you do?

SCENARIO 2

You are a 16-year-old girl. It's Saturday night and you have been invited to a party on the beach. You ask your parents to drop you off at a friend's house to do some study. As soon as they drop you off, your friend's sister grabs her keys and a six-pack of beer and tells you to hurry up. Before you jump into the back of her car, your friend's sister ushers you to the bedroom where you are confronted with two lines of cocaine on a compact mirror. She says, 'let's have some real fun'. What do you do?

SCENARIO 3

You are a 15-year-old boy. It's a hot summer's day and your parents are away for the weekend. You invite a few of your mates around for a swim. One of your mates brings his older brother, who has brought a slab of beer. You all start to relax and unwind around the pool, listening to music. Your mate's older brother has had several drinks and thinks it would be a great idea to set up some ladders and jump off them into the pool. You are sober. What do you do?

SCENARIO 4

You are a 16-year-old girl and you are at a friend's party. You brought your own alcohol but by 9 p.m. you have run out! You recognise an older male student from school who approaches and offers you a can of premixed drink. The can appears full but it is already open. What do you do?

SCENARIO 5

You are a 17-year-old girl. You are at a friend's house and it is late. You ordered an Uber but it still hasn't showed. It is 1.30 a.m. and you really want to go home. You know that the walk home would only take 20 minutes. What do you do?

WORKSHEET 1.10 TV COMMERCIAL

SB
Pages 44

Advertising executives use persuasive writing in commercials to either sell a product or to influence or alter decisions made by an audience. Most TV commercials are designed to have an impact on viewers' emotions.

Parties and gatherings can be lots of fun. But, whether hosting your own party or invited to a friend's gathering, it's important to ensure that you and your friends are safe. Read the 'Partying safely – tips for teenagers' fact sheet at the Victorian Government Better Health Channel website (you can Google 'Better Health Channel partying safely tips for teenagers' to find this).

1 In groups of five or six, write a 30- to 60-second informative commercial script titled 'Tips to partying safely'. The script is to include multiple characters. You will have 15 to 20 minutes to write the script and 10 minutes to practise performing the TV commercial. Brainstorm ideas for the commercial in the space below, and then write the script in your notebooks or on a computer.

2 Each group will then take a turn presenting their commercial.

WORKSHEET 1.11 BLOOD-BORNE VIRUS SAFETY

Pages 38–9

Blood-borne viruses (BBVs) are viruses that are found in blood or bodily fluids. BBVs can be transmitted from person to person via the sharing of needles associated with drug use or by risky behaviour, such as unprotected sex. Risky behaviour is often associated with alcohol and other drug use.

iStock.com/ddukang

Using the following website, research BBVs.

Weblink
Get the facts: Blood safe

In groups of four, choose a topic below to investigate and report your findings to the rest of the group.

1 Blood-borne viruses – Hepatis B, Hepatitis C, HIV and AIDS
2 Blood-borne viruses – Sexually transmitted infections
3 Blood-borne viruses – Injecting drugs
4 Blood-borne viruses – Body art

Topic selected:

Notes:

9780170465540

WORKSHEET 1.12 EMERGENCY SITUATIONS

Pages 46–8

Very likely, at some stage in your life, you may find yourself faced with an emergency or injury. Knowing what to do and being able to respond quickly and efficiently could make a difference and even save a life. When an emergency occurs, remember to:

1 Take a deep breath, count to 10 and tell yourself that you can handle this situation.

2 Follow DRSABCD steps:

   Danger

   Response

   Send for help

   Airway

   Breathing

   CPR

   Defibrillator

3 Call an ambulance if required.

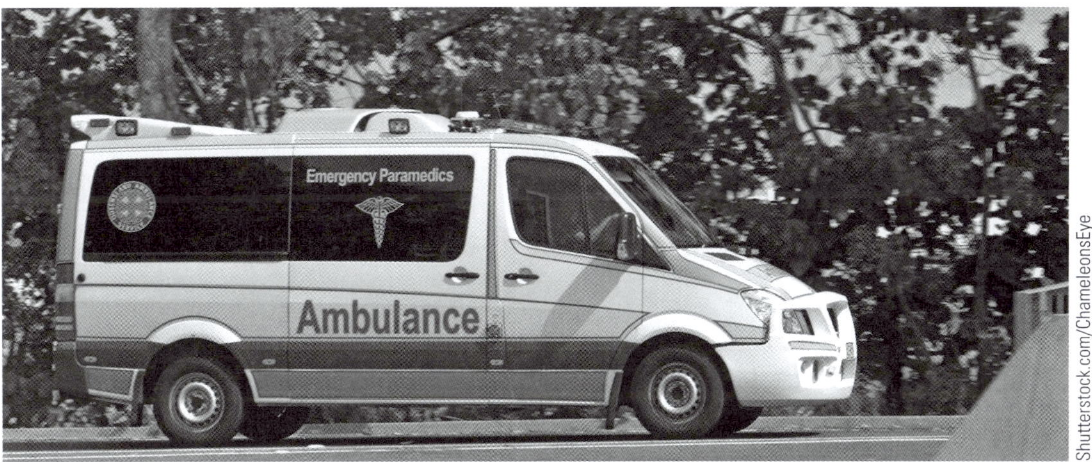

Shutterstock.com/ChameleonsEye

4 Decide what you would do in each of the following emergency situations, and which of the **DRSABCD** steps you would follow.

WORKSHEET 1.12 CONTINUED

Emergency situation	Your response
You are at a friend's party. You have not been drinking but you notice your friend Chase has been carrying around a 5-litre box of cask wine, and it is nearly empty. You watch him stagger down some steps, accidentally tripping and hitting his head hard on the concrete floor. Chase is conscious but you notice blood oozing from a deep cut to the back of his head. No adults are present.	
A partygoer who you haven't met before has suddenly collapsed, and you have no idea why. She is not responding but is breathing.	
You are waiting for a tram and notice a group of girls inhaling something from a paper bag at the other end of the platform. One of the girls is staggering towards you. She appears disorientated and confused. She is headed towards the tracks as the tram is approaching.	
You are playing a card game with some of your friends. The loser of each game has to drink a shot of tequilla. One of the boys is really drunk, stumbles to the bathroom and vomits violently. You find him unconscious on the bathroom floor.	

9780170465540

WORKSHEET 1.13 PERFORMANCE-ENHANCING DRUGS IN SPORT

Page 50

The men's 100-metre Olympic final at the 1988 Seoul Olympics was once dubbed 'the dirtiest race in history'. The winner, American Ben Johnson, was stripped of his gold medal two days after setting a world record time after he tested positive for steroid use. There had been much speculation that others who ran that infamous race were also using performance-enhancing drugs.

Getty Images/Mike Powell

Ben Johnson was disqualified from the 1988 Seoul Olympics for doping.

In 2012, at the London Olympics, the women's 1500-metre final will also be remembered as one of the dirtiest races in history. Of the 13 women who raced on 10 August 2012, six have been suspended for doping at some point in their professional careers.

CASE STUDY

Doping and the women's 1500 metres at London 2012

The results of the women's 1500 metres at the 2012 London Olympic Games are mired in scandal after six out of the first nine athletes to finish the race have been linked to banned performance-enhancing drugs, either before the race or in the years that followed.

The first nine finishers, their ages at the time, the country they represented, their ranking and race times and any doping history are listed below.

1 Asli Çakir Alptekin, 27, Turkey, first (gold) in 4:10.23

Çakir Alptekin had already received a two-year ban in 2004. In 2015, she was given an eight-year ban for biological passport violations and was stripped of all her results from 2010 onwards, including her gold medal from the London Olympics. The time on her ban was reduced but she was again banned – this time for life – in 2017.

2 Gamze Bulut, 20, Turkey, second (silver) in 4:10.40

Bulut set her PB or performance best of 4:01.18. Previously her PB had been 4:18.23 from a race in July 2011. Bulut has also had irregularities in her biological passport and received a four-year ban and had her medals and records stripped from 2011 to 2016.

3 Maryam Yusuf Jamal, 27, Bahrain, third (bronze) in 4:10.74

Jamal was born in Ethiopia but ended up running for Bahrain after seeking political asylum. She has never failed a drug test. In 2017, she was given the gold medal for the London race since both Çakir Alptekin and Bulut's medals had been stripped. Jamal had also won gold at the 2007 and 2009 World Championships.

4 Tatyana Tomashova, 37, Russia, fourth in 4:10.90

Tomashova had previously won an Olympic silver medal in the 1500 metres at Athens in 2004. She was suspended in 2008 for 'fraudulent substitution of urine' and received a two-year ban. Tomashova was reallocated the silver medal after Çakir Alptekin and Bulut lost their medals.

5 Abeba Aregawi, 22, Ethiopia, fifth in 4:11.03

Aregawi originally represented Ethiopia but began running for Sweden in 2013 and won a gold medal at the 2013 World Championships. In February 2016, Aregawi tested positive for meldonium, which was added to the banned substances list in the month before. She was suspended from competition until her ban was overturned. There was insufficient evidence regarding how long it would take for the substance to leave the body.

6 Shannon Rowbury, 27, United States, sixth in 4:11.26

Rowbury has never failed a drug test. She won bronze at the 2009 World Championships and placed fourth at the Brazil Olympic Games in 2016. She set the American record for her time of 3:56.29 in the 1500 metres in 2015, although this was broken by another athlete in 2019.

7 Natallia Kareiva, 26, Belarus, seventh in 4:11.58

Kareiva received a two-year ban in 2014 when her biological passport showed abnormalities and she was disqualified from all races from 2010 to 2014, including the 2012 Olympics.

8 Lucia Klocová, 28, Slovakia, eighth in 4:12.64

Klocová has never failed a drug test. She had previously run in the semi-finals of the 800 metres at the Athens and Beijing Olympics.

9 Yekaterina Kostetskaya, 25, Russia, ninth in 4:12.90

Kostetskaya was given a two-year ban in 2014 after her biological passport showed abnormalities and she was found guilty of 'use/attempted use of a prohibited substance/method'. She was disqualified from both the 2011 World Championships (where she placed fifth in the 800 metres) and the 2012 Olympics.

Photo Run/Victor Sailer

Has never failed a drug test

Has been implicated in doping

9780170465540

Complete the following questions using information from the article.

1 The article refers to the term 'biological passport' when discussing athletes who have been caught doping. Explain what this means.

____

____

____

2 Of the nine athletes who competed in the 1500-metre race, exactly how many of these athletes had never failed a drug test?

____

3 Describe what happened to first-placed Turkish athlete Asli Çakir Alptekin after her biological passport appeared to be have been violated?

____

____

There have been a number of athletes who have chosen to cheat their way to success, and many have been caught along the way. Cheating is in opposition to the concept of 'fair play' in sporting competitions. Using the article, your student book and the internet, answer the following questions.

4 Describe what you consider to be the main issue affecting the integrity of professional sport.

____

____

____

____

____

5 In your own words, define the term 'fair play'.

____

____

____

6 Athletes dope in many ways. Identify and investigate three banned substances used to enhance sporting performance.

____

____

____

WORKSHEET 1.13 CONTINUED

7 Read the following scenario and answer the questions below.

Trey (15 years) refused to participate in school swimming activities as he felt embarrassed with his body shape and the way he looks. He is concerned that he doesn't have muscles like some of the other boys in his year and he lacks confidence in front of others. Trey has started using performance-enhancing drugs to help him develop more muscle tone. He told his best friend Jamie who is concerned about Trey's decision and suggests that he sees a doctor for additional advice. Trey refuses and says he knows what he is doing.

a Identify the performance-enhancing drug Trey may have taken to build muscles.

____

b Discuss one positive and one negative way in which Trey could deal with this situation.

____

____

c Discuss one positive and one negative way in which Jaime could deal with this situation.

____

____

d If Trey was your friend, summarise the advice you would offer.

____

____

____

Getty Images/Cameron Spencer

British Olympic silver medallist Chijindu Ujah tested positive for two banned substances after winning a silver medal in the 4 x 100-metre relay at the 2020 Tokyo Olympic Games.

WORKSHEET 1.14 TYPES OF DOPING DRUGS

Page 51

Many athletes have been caught taking illegal drugs to enhance their sporting performance. In groups of four, choose one illegal substance to investigate further. Answer the following four focus questions based on the illegal substance you have chosen to investigate.

Drug chosen ______________________________

1 Identify which sports the illegal substance is used, and why.

____
____
____
____

2 Investigate how this substance enhances sporting performance.

____
____
____
____
____
____
____
____

3 Determine the health risks associated with using this substance.

____
____
____
____
____
____
____
____

4 Name athletes who have been caught using this drug to enhance performance. Discuss the penalties they received as a result of their illegal drug use.

examine to investigate, inspect or scrutinise; inquire or search into; consider or discuss an argument or concept in a way that uncovers the assumptions and interrelationships of the issue

AC

5 **Examine** alternative healthier options that can be used to enhance performance in the same way as the substance you are investigating.

9780170465540

CHAPTER 1 REVIEW

Reflect upon your learning in this chapter by completing the following sentences.

1 Information I found useful includes …

2 I can use my new knowledge to …

3 It was surprising that …

4 I have changed my opinion about …

5 It may have been good to also consider …

2

NUTRITION AND SUSTAINABILITY

Shutterstock.com/Nicetoseeya

WORKSHEET 2.1 FRUIT AND VEGETABLES

Page 56

Analyse the graph below, which shows how much fruit and vegetables Australians aged 2 and over eat each day.

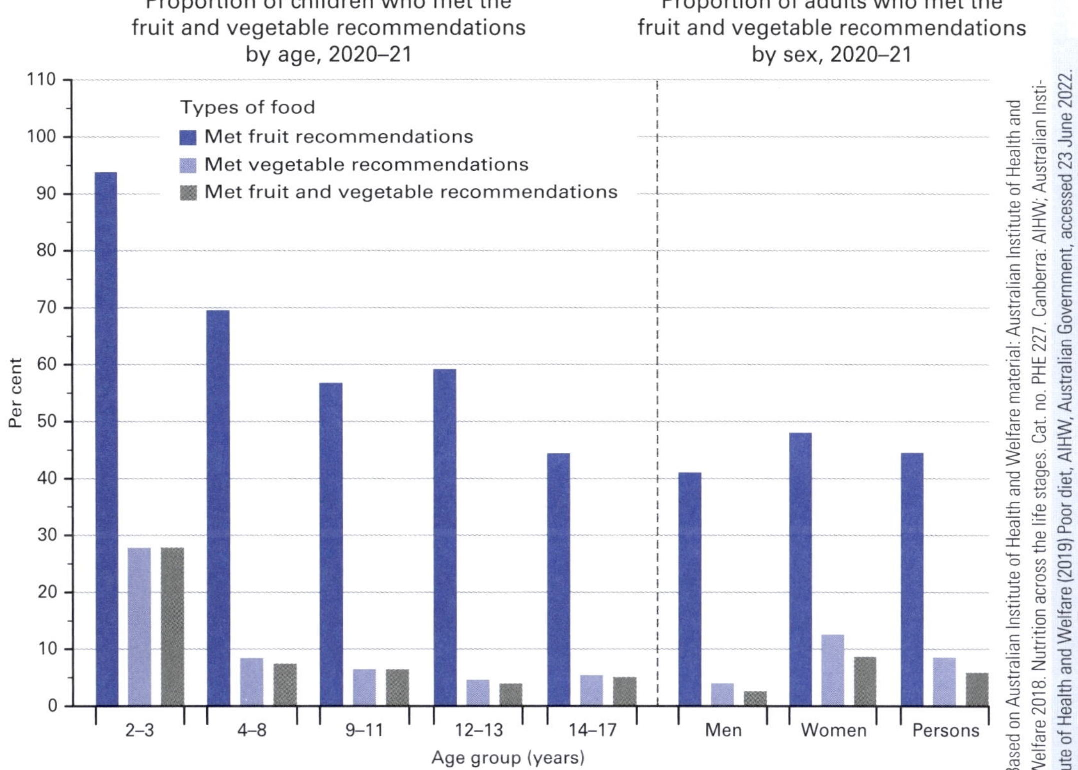

Based on Australian Institute of Health and Welfare material: Australian Institute of Health and Welfare 2018. Nutrition across the life stages. Cat. no. PHE 227. Canberra: AIHW; Australian Institute of Health and Welfare (2019) Poor diet, AIHW, Australian Government, accessed 23 June 2022.

1 Determine which age range of children, and of adults, are consuming the most amount of vegetables per day.

____

____

2 Determine which age range of children, and of adults, are consuming the most amount of fruit per day.

____

____

3 **Hypothesise** some reasons for such a big discrepancy between the levels of fruit and vegetable consumption.

____

____

____

____

hypothesise
to formulate a supposition to account for known facts or observed occurrences; conjecture, theorise, speculate; especially on uncertain or tentative grounds

WORKSHEET 2.1 CONTINUED

4 **Describe** the trend of the levels of consumption of fruit and vegetables across the age groups. What do you notice? Why is this the case?

5 Which age ranges are meeting the recommended daily intake of fruit or vegetables in Australia? Why do you think so many people are not eating the recommended amounts of fruit and vegetables per day?

describe to give an account of characteristics or features

AC

6 **Propose** some strategies that people could use to ensure they are getting the recommended amounts of fruit and vegetables each day.

propose to put forward (e.g. a point of view, idea, argument, suggestion) for consideration or action

AC

WORKSHEET 2.2 FRUIT VERSUS JUICE

SB
Page 66

Using the table below, **compare** the nutrient value of an apple and a glass of apple juice.

compare
to observe or note how things are similar or different

Shutterstock.com/Africa Studio

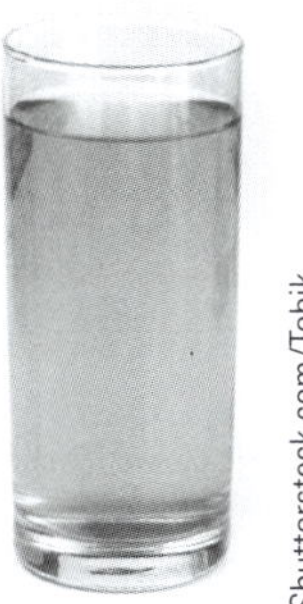

Shutterstock.com/Tobik

	Apple	Apple juice
Size	approximately 200 g	up to 650 mL
Fibre	3–4 g	0–1 g
Energy	approximately 300 kJ	up to 1300 kJ
Time to eat	5–10 minutes	2–5 minutes

Consider the following facts about juices:

- Although juice drinks can be a useful source of nutrients, they have fewer nutrients than whole fruit, and they often have as much sugar as soft drinks. Consuming these drinks provides you with the sugar and energy of several pieces of fruit, but not necessarily the nutrient value of several pieces of fruit (even when they are marketed as having 'no added sugar').
- Large (400 mL or more) juice drinks can contain more than 50 g of sugar (the daily recommended amount of sugar for adults is 90 g).
- They can contain well over 100 kJ of energy (the daily recommended amount of energy for adults is 8700 kJ).
- Juice drinks contain no protein and very little fibre (it is removed when the fruit is juiced). This contributes to a lack of satiety (feeling full).
- The time and effort to consume a large juice drink is quite small compared to eating whole fruit.
- Some juices from 'juice bars' contain added caffeine, and it is usually not clear exactly how much.
- After drinking a large juice, some people develop the false perception that they have had their 'fix of vegies/fruit' for the day, so it doesn't matter what else is consumed for the day.

WORKSHEET 2.2 CONTINUED

1 Write a paragraph that outlines how you think juices can be included as part of a healthy diet. Some suggestions you may want to consider include using water to quench thirst, eating whole fruit, not relying on the supposed nutrients in juices to replace complete and balanced meals, how often and what size juices are consumed, treating the energy value of juices as a snack or mini-meal instead of a drink, and how to overcome the social appeal of purchasing large cups of juice from commercial 'juice bars'.

____

____

____

____

____

____

____

____

____

____

____

____

2 Using the information in the table, create a graph below that compares the energy and fibre of apple juice versus that of a whole apple.

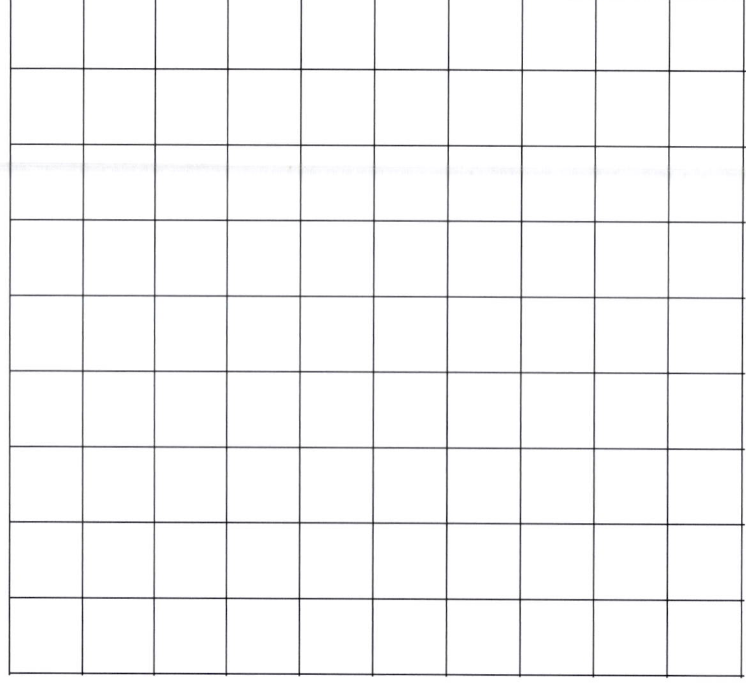

9780170465540

WORKSHEET 2.3 UPSIZING

SB Page 71

Nutritionists and dieticians refer to increases in size of portions as 'portion creep' or 'portion distortion'; the size of drinks, snacks and meals has grown considerably over time, usually as a means for companies to generate more sales by implying that their products represent better value (more food or drink for less money).

Read the table below, and note the changes in the energy content of these common drinks, snacks and meals from 20 or more years ago to the present day. The increase in energy is completely due to the increase in portion sizes (i.e. the size of the cup, packet or food item).

	Energy (kJ) 20+ years ago	Energy (kJ) present-day	% increase
Slurpee	244	1410	477
Coke	450	1080	
Fruit juice	285	855	
Movie popcorn	1265	3455	
Muffin	810	2580	
Cookie	231	1155	
Pizza slice	2100	3570	
Cheeseburger	1398	2478	
Fries	925	1670	

1 Use the formula below for each food item to calculate exactly how much the energy content of each food item has increased, and record your answers in the '% increase' column in the table above.

$$\frac{(\text{Energy present-day} - \text{Energy 20+ years ago})}{\text{Energy 20+ years ago}} \times 100\% = \%\ \text{increase}$$

For example, Slurpee: $\frac{(1410 - 244)}{244} \times 100 = 477\%$

WORKSHEET 2.3 CONTINUED

compare to observe or note how things are similar or different

2 Create a bar graph to **compare** the % increases in the space below.

3 Suggest some commonalities between these types of foods and drinks. Hint: where would they fit in the Australian 24-Hour Movement Guidelines for Children and Young People (5–17 years)?

____

____

____

4 Propose two reasons why portion sizes of these types of foods and drinks have increased so much in 20 years.

____

____

____

5 Describe some strategies you use to avoid consuming too-large portions of these types of foods.

____

____

____

____

WORKSHEET 2.4 TIPS FOR HEALTHY MEALS

SB
Page 71

Below are several tips to have healthier meals:

- Use reduced-fat versions of dairy products.
- Choose unsaturated fats over saturated fats. This information will always be included in the nutritional information on food packaging.
- Never skip breakfast, as it means you will be more likely to give in to the temptation of discretionary foods later in the day.
- If having fast food for a meal, choose bread-based foods such as wraps, sandwiches or kebabs instead of pastry or deep-fried options. Limit your use of sauces, and only upsize if it is a salad option.
- Choose lean meats over processed or cured meats.
- Trim skin and fat from meat and poultry.
- Keep your thirst satisfied by sipping water regularly throughout the day, and have water before and during your main meals.
- Make sure you choose a range of different coloured vegetables to increase your chances of eating a wide variety of vitamins and minerals.
- Choose wholemeal over white bread and wholegrain cereals over sugary cereals.
- Go for regulated sizes when snacking on discretionary foods. For example, an ice-cream on a stick is a limited amount while ice cream scooped out of a tub may be a larger serving.
- Eat slowly, without distractions such as the TV, put your cutlery down between mouthfuls when you're chewing, and sip water between mouthfuls to slow your pace. Concentrate on how your meal looks, smells, tastes and feels in your mouth before you swallow. This will help you to enjoy the food more and eat smaller portions.

1 Identify some tips from the list above that you have tried to incorporate into your eating habits. What are some challenges in doing this consistently?

WORKSHEET 2.4 CONTINUED

2 Select one or two tips listed above that you haven't tried. Develop a strategy to support you in introducing them into your lifestyle and making them a part of your healthy diet.

3 Can you come up with any more tips that you can share with your class? Create a list of five and then share your ideas.

4 From the list of tips above, rank the three tips that would be easiest to consistently implement, and rank the three tips that would be hardest to consistently implement.

Answer space:

Easiest	Hardest
1.	1.
2.	2.
3.	3.

9780170465540

WORKSHEET 2.5 A HUNGER LOG

Page 77

A hunger log can be a useful tool to help understand how your stomach feels and the times of day when you truly feel hungry; this can be useful to stop you over-eating. Many Australians have portions, meals and snacks that are unnecessarily large, leading to the consumption of too much energy, and thus weight gain. It also means a lot of food is wasted by throwing away what is not eaten. It is normal to feel quite hungry a few times a day, particularly if you are active. This is your body telling you it is time to eat.

Fill in a hunger log using the template below, shading in your level of hunger from 1 to 10 throughout the day. Do this over a few days.

For a healthy diet, try to stick to the following guidelines:

- Wait until your hunger is at 1, 2 or 3 before you eat.
- Stop eating when you reach 5 (if you eat too quickly, you may reach 6 or higher before you realise – eat slowly!).
- Eating until you are at 7 or above may lead to weight gain in the future.

Hunger rating scale

1. I'm absolutely ravenous
2. I feel really hungry
3. I'm running on empty
4. I'm a bit peckish
5. I feel quite satisfied
6. I probably didn't need that last mouthful
7. I'm rather full
8. I'm uncomfortably full
9. I'm really overstuffed
10. I'm about to burst and I feel sick

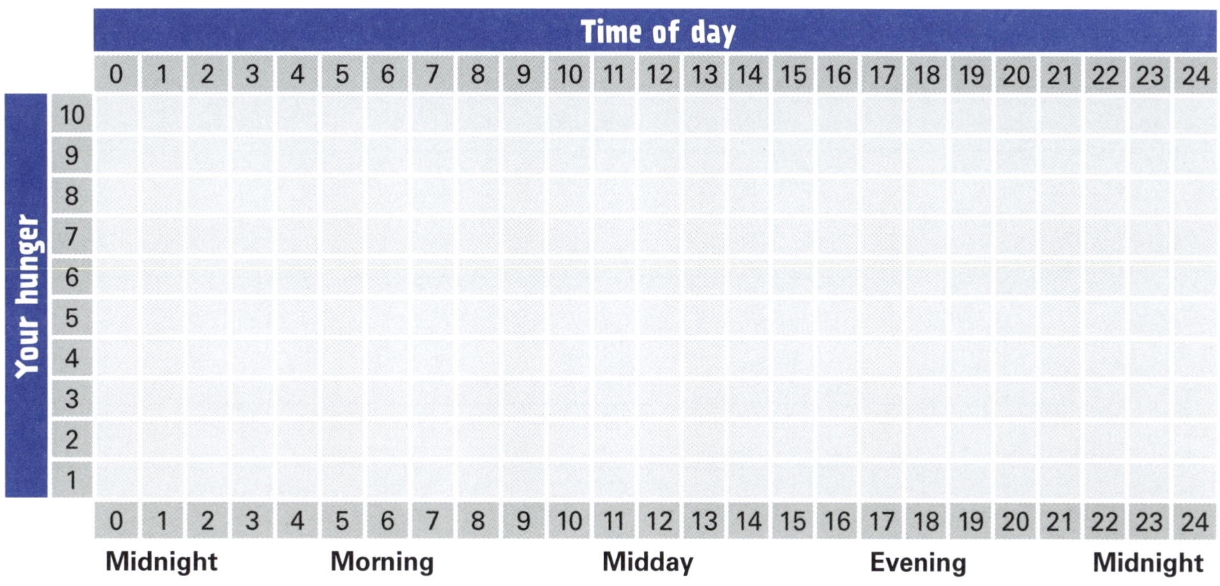

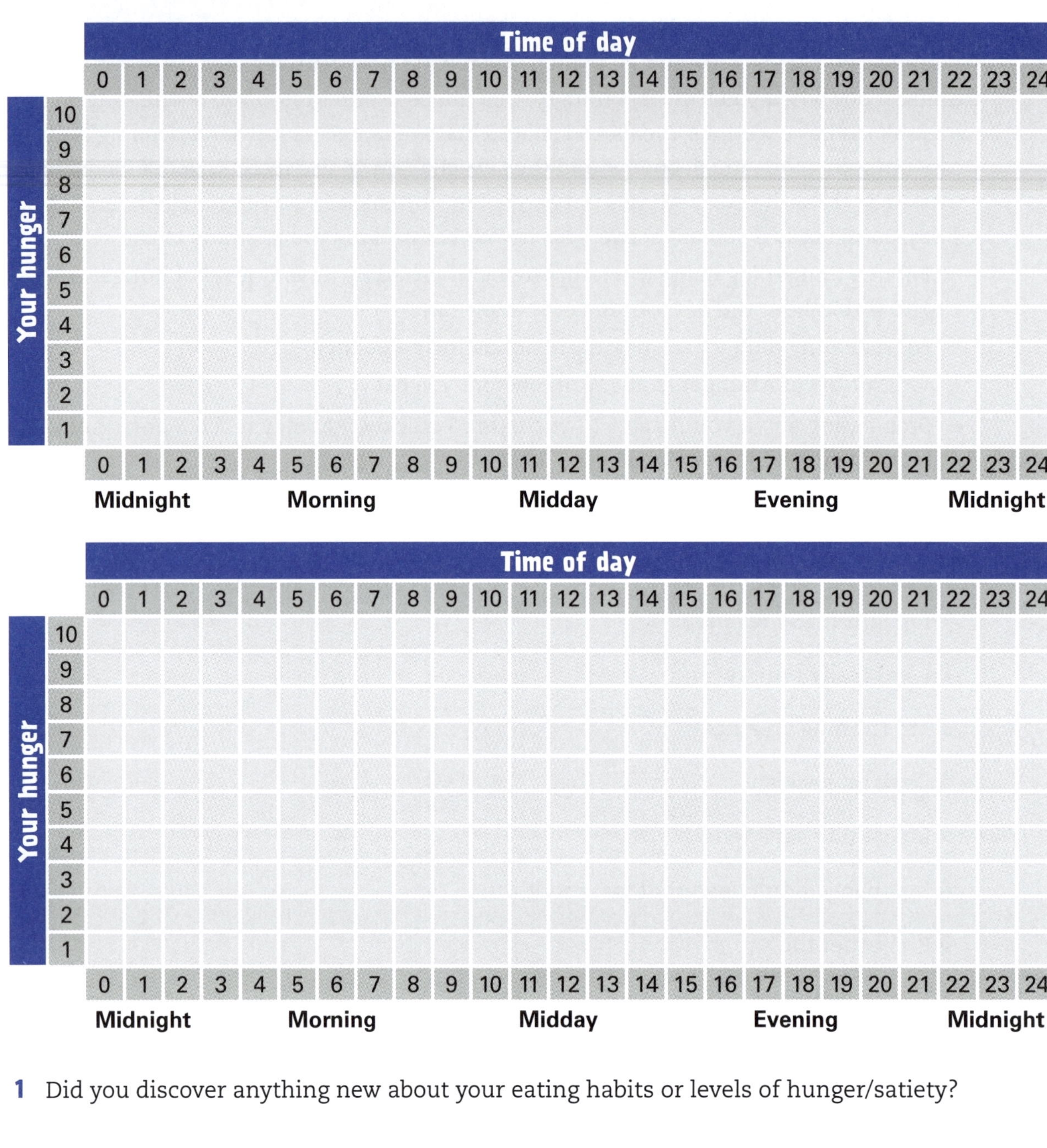

1 Did you discover anything new about your eating habits or levels of hunger/satiety?

2 Identify the times of day when you are hungriest. Now that you know this, develop a strategy to make sure that you always have a healthy snack or meal available to you at these times.

3 Evaluate how well were you able to stick to the suggested guidelines. What was challenging about this?

4 Suggest why you think it is important to wait until your hunger is at a rating of 1, 2 or 3 before eating again.

5 Propose why you think it is important to stop eating when a rating of 5 is reached, instead of eating until you are overfull.

WORKSHEET 2.6 OBESITY

SB Page 79

This graph shows the level of obesity (not just overweight) among adults and children in Australia from 1980 to 2018.

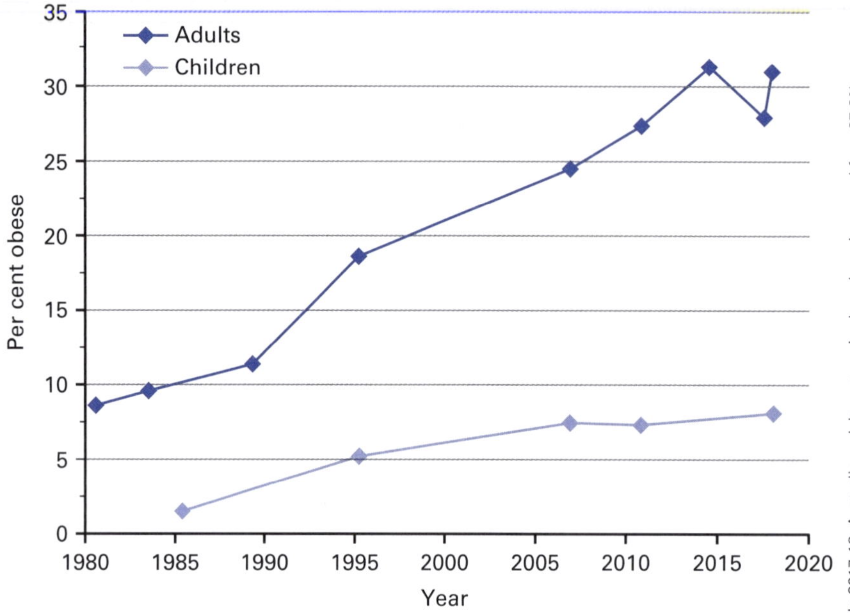

In 2017-18, Australian adults categorised as obese increased from 27.9% (in 2014-15) to 31.3%. Of children aged 5-17 years 8.1% were obese. The rates were similar for boys and girls and this has remained stable over the previous 10 years.

Based on Australian Bureau of Statistics data, National Health Survey: First results, 2017-18 financial year, https://www.abs.gov.au/statistics/health/health-conditions-and-risks/national-health-survey-first-results/2017-18, ABS website, accessed 15 August 2022

1 Identify the lowest percentage of adults and children with obesity in Australia, and when this occurred.

____

____

2 Identify the highest percentage of adults and children with obesity in Australia, and when this occurred.

____

____

describe to give an account of characteristics or features

3 **Describe** the trend over time of adults with obesity in Australia. Can you predict what has happened since 2020? Why do you think this is?

____

____

____

____

____

____

____

____

9780170465540

4 Hypothesise some reasons behind the trends depicted in the graph.

5 Describe some additional health consequences that obesity can cause or contribute to.

6 What strategies have been tried to change these trends? Why haven't they worked? What could be tried instead?

WORKSHEET 2.7 OBESITY INTERVENTIONS

Pages 81

To really curb the obesity epidemic, governments may need to employ 'population-level interventions' such as:

- Tax unhealthy foods
- Subsidise healthy foods such as fruits and vegetables
- Limit the availability of unhealthy foods, for example in schools
- Limit the advertising of unhealthy foods, especially to children

explain to make an idea or situation plain or clear by describing it in more detail or revealing relevant facts; give an account; provide additional information

1 **Explain** what you think is meant by the term 'population-level interventions'.

____

____

____

____

2 Adding a tax to alcohol and tobacco to increase their price has been used for decades as an effective strategy to limit the consumption of these products. State if you think a tax on unhealthy foods would be effective, and justify why or why not.

____

____

____

____

3 If you were buying a snack or a meal, and the junk food option was noticeably more expensive than the healthier option, would this influence your decision? Explain why or why not.

____

____

____

____

determine establish, conclude or ascertain after consideration, observation, investigation or calculation; decide or come to a resolution

4 Of the four suggested interventions listed above, **determine** which one would be most effective and why.

____

____

____

9780170465540

SB
Pages 85

WORKSHEET 2.8 OBESITY NORMALISED

CASE STUDY

Obesity has become the new normal but it's still a health risk

Nike's London store recently introduced a plus-sized mannequin to display its active clothing range which goes up to a size 32.

The mannequin triggered a cascade of responses ranging from outrage to celebration. One side argues that the mannequin normalises obesity and leads obese people to feel that they are healthy when in fact they are not.

The other side argues the representations are inclusive, combat fat stigma and encourage fat women to exercise. Both arguments have some merit.

Shutterstock.com/JessicaGirvan

Nike introduced plus-size mannequins in its London flagship store.

The representations of bodies we see around us – including shop mannequins – affect the way we calibrate our sense of what is normal and acceptable. And obesity is indeed associated with a greater risk of heart disease, stroke, type 2 diabetes and early death.

It is possible to be metabolically healthy and fat. But even metabolically healthy obese people may still have a shorter life expectancy than their lean peers.

On the other hand, exercise is almost universally beneficial, and people of all shapes and sizes should be encouraged to participate.

Overweight and obesity have become the new normal

Based on body mass index (BMI), about two-thirds of Australian adults and one-quarter of kids are overweight or obese. While this proportion has flattened out for children in the last 20 years, it continues to rise for adults.

There is strong evidence parents consistently misjudge the weight status of their children because they see more and more fat kids.

The same is true for adults: a recent study from the United Kingdom found 55% of overweight men and 31% of overweight women considered their weight to be in the healthy range.

I would guess the Nike mannequin is close to 100 kg, with a BMI maybe in the low 30s, well into the obese category.

But given the average female shop mannequin has a BMI of about 17, there are probably at least 10 times as many Australian women like the plus-size mannequins than like the usual minus-size variety.

Obesity is not a lifestyle choice like smoking

Obesity is necessarily the result of behaviours – eating too much, exercising too little – albeit heavily constrained by genetic predispositions, and social and economic pressures.

But unlike, say, smoking, being fat is also part of what a person is: most people who are fat have usually been fat for a long time. It's not something a person has complete control over.

Divergent paths into fat and lean start very young, and once you're on the obesity train it's hard to get off.

While it is possible to 'give up obesity', for many it can be a very hard road, involving a lifelong struggle with hunger and recidivism.

Empowering vs shaming

Weblink
Measure Up campaign

Anti-obesity campaigns that are built on disgust, fear or shame – such as **Measure Up** – have been criticised as being stigmatising, ethically problematic and ineffective.

There has, to my knowledge, been no high-quality research comparing the actual effectiveness of shaming versus empowering anti-obesity, or pro-physical activity, campaigns.

However, a number of studies show, unsurprisingly, that obese and inactive people prefer empowering campaigns, find them more motivating and less stigmatising.

Health risks of obesity

It has been argued one can be 'fit and healthy at any size': that an obese person can be as fit and healthy as a lean person.

Depending on definitions, about 25–50% of obese people have 'metabolically healthy obesity' – normal levels of inflammation, blood sugar, insulin, blood fats, and blood pressure. Other than being obese, these people appear healthy.

But obese people – fit or unfit, active or not – remain on average at greater risk of heart disease, diabetes and early death than lean people with similar behaviours.

Similarly, the claim that people can be both fit and fat, and that fit, fat people are at less risk than unfit, lean people depends on how we define fitness and fatness.

One study, for example, might compare overweight people in the top 20% of fitness with lean people in the bottom 20%. Because there are modest differences in fatness and big differences in fitness, fat people are much more likely to have a similar risk to lean people.

But if another study compares obese people in the top 50% of fitness to lean people in the bottom 50%, the fatter people will be much less healthy.

What is certain is that whoever you are, exercise will almost certainly improve your health.

The Nike mannequin controversy is a morality tale of how we navigate between the devil of normalising obesity and the deep blue sea of excluding obese people from the world of exercise.

Obesity has been called both a disability and a disease, and just another way of being in the world. The reality is that for most people, it's something in between.

'Obesity has become the new normal but it's still a health risk' by Tim Olds, Professor of Health Sciences, University of South Australia, *The Conversation,* https://theconversation.com/obesity-has-become-the-new-normal-but-its-still-a-health-risk-118829

WORKSHEET 2.8 CONTINUED

1 Identify an example from the article that support the notion that overweight and obesity have become the new normal. Can you suggest any examples of your own that you have come across in your daily life?

2 Society has reacted with 'outrage and celebration' about the normalisation of obesity. Assert where you stand on this issue.

3 What does it mean to be 'overweight but metabolically healthy'? The author states this is possible, to a degree. Outline the evidence suggested in the article that work for and against this statement. Declare whether you agree with this or not, and justify why.

4 The author states that 'obesity is not a lifestyle choice like smoking'. What justification does the author use to support this statement? Do you agree?

WORKSHEET 2.8 CONTINUED

5 The author is critical of some anti-obesity health promotion campaigns. Summarise the basis of his criticism.

6 Propose how you would go about designing an anti-obesity health promotion campaign which take into consideration the criticism outlined in this article.

WORKSHEET 2.9 ROLE-PLAY

Page 85

Select one of the scenarios below, and with a classmate, act out each role. Use your nutritional knowledge to provide the healthiest advice and achieve the best outcome.

SCENARIO 1

You meet your friends at the cinema to see a movie. One of your friends decides to upsize their snacks and drink. You make a healthier choice and persuade your friend to do so too.

SCENARIO 2

One of your good friends is trying out a new fad diet that they heard was popular with some celebrities and sports stars. You know that this diet won't give your friend the nutrients they need, and the extreme lack of kilojoules is already making them lethargic and irritable. How could you try to persuade your friend to not use this crazy diet?

SCENARIO 3

Your parents have taken you and your little brother/sister shopping at the supermarket for the weekly groceries. Your brother/sister is being a real pain, and nagging and whinging about a new energy-dense and nutrient-poor food item that they saw advertised on TV. Explain why your parents shouldn't buy it, and what some healthier alternatives might be.

SCENARIO 4

Your brother/sister likes to sleep in so they never eat breakfast, although sometimes they have a sugary snack on the way to school. How can you convince them to develop a better routine in the morning?

WORKSHEET 2.10 OUR FOOD WASTE

Pages 85

1 Take note of the following statistics about Australian household food waste. Shade in each illustration to represent these facts and figures. How many shopping bags do you need to shade? How much of the bin?

- On average, Australian households throw out 20% of the food they purchase.

- Food makes up 40% of the average household rubbish bin.

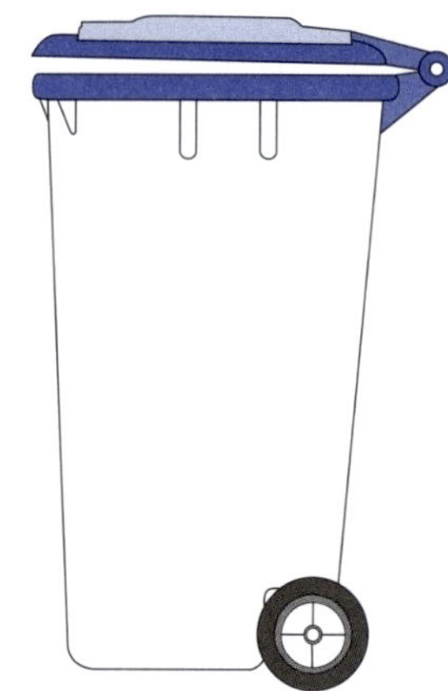

- Australians throw out about $8 billion of edible food as household waste every year. This figure can be broken down into:
  - $2.67 billion of fresh food = 33%
  - $2.18 billion of leftovers = 27%
  - $1.17 billion of packaged and long-life products = 15%
  - $727 million of drinks = 9%
  - $727 million of frozen food = 9%
  - $566 million of takeaways = 7%

2 Shade the pie chart below and label each segment to best represent the different types of waste listed above.

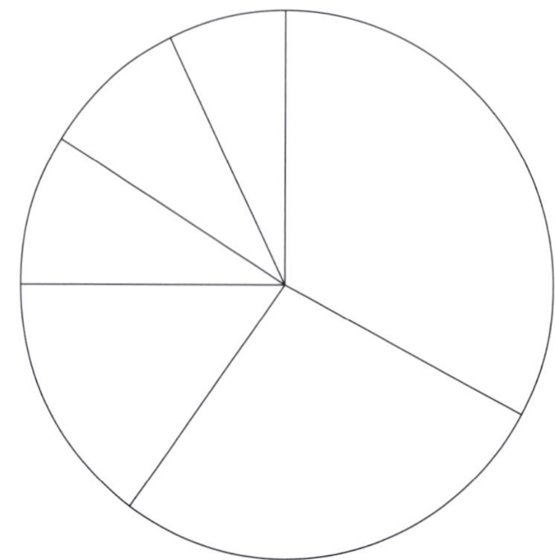

9780170465540

3 Common reasons why people throw out so much food as waste include:

- Cooking too much food
- Spontaneously buying takeaway meals instead of cooking the food already at home
- Not knowing how to utilise leftovers
- Mistakenly throwing food out before the use-by/best before date
- Not checking the pantry or fridge properly before shopping for groceries
- Purchasing too many food or drink products due to not sticking to a shopping list
- Shopping when already feeling hungry, leading to buying too much

Propose three other reasons:

____

____

____

4 Using the table below, rank the list above in order of which reasons are more likely to cause the food to be thrown out (you can include your own ideas). 1 = most likely, 10 = least likely.

Rank	Reason
1	
2	
3	
4	
5	
6	
7	
8	
9	
10	

5 For each of your top three ranked reasons, **describe** a strategy to minimise the amount of food thrown out by households.

describe to give an account of characteristics or features

____

____

____

____

____

WORKSHEET 2.11 SUSTAINABILITY PROJECTS

Pages 89–91

Use the table below to develop the plan for your sustainability project, as discussed in Chapter 2 of the student book.

Project title	
Goals (What do we hope to achieve? What will the finished product look like? How will it improve our school and the environment in general?)	
Stakeholders (Who do we need to liaise with or involve, either within or outside of the school? What role do they play?)	
Consultation (What is the process of gaining approval and ensuring input from relevant stakeholders? How do we satisfy the requirements of the process?)	
Communication (How do we promote our project and keep the school and wider community aware of our progress and outcomes?)	
Resources (What materials and equipment do we need? How will it be appropriated? How much funding will we need and from where do we obtain the funding?)	
Timeline (When do we hope to complete our project? Consider different stages such as planning, gaining approvals, fundraising, gathering resources, creating/building, implementation and use, and review of the project.)	
Review (Consider your goals and the process of working towards those goals: What were you proud of? What would you do differently next time? Was the project a success and how do you know if it was?)	

WORKSHEET 2.12 AGREE OR DISAGREE

Pages 92

1 How strongly do you agree with the statements below? For each statement, place its number where you think it best fits on the continuum. There are no right or wrong answers.

Be prepared to **justify** your response in a class discussion.

justify to show how an argument or conclusion is right or reasonable

a If the government subsidised (reduced the cost of) healthy food options in school cafeterias, it would improve the diet of children and adolescents.

b Social media influencers recommending food and drink products are a reliable source of nutritional information and dietary advice.

c Energy drinks have a 'recommended daily limit' of two cans or bottles, but there is no law to stop anyone buying and drinking more than this. Should there be laws introduced to change this?

d People who are within a healthy weight range must have a good diet.

e Discretionary food items are best eaten as occasional treats.

f We can grow enough food to feed everyone on the planet if we equitably distributed it and valued it enough not to waste it.

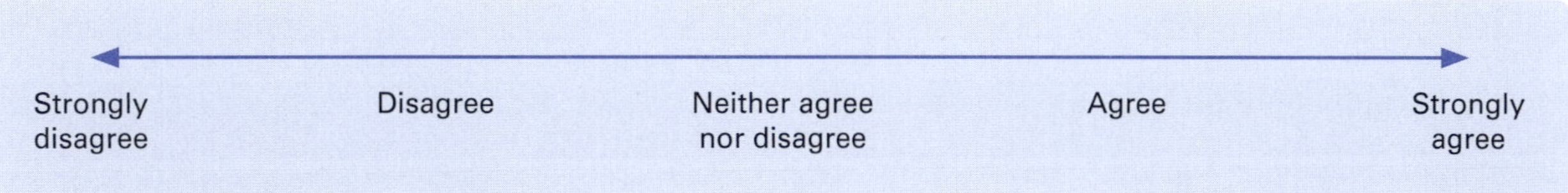

2 Discuss your responses with a partner. Now re-evaluate your ratings. Would you change your response to anything?

CHAPTER 2 REVIEW

Reflect upon your learning in this chapter by completing the following sentences.

1 I developed a particular strength in ...

2 I was surprised to discover ...

3 Information I found useful includes ...

4 I am looking forward to ...

5 I hope more people get the chance to ...

6 I can see my lifestyle changing by ...

9780170465540

3

HEALTHY PEOPLE, HEALTHY COMMUNITIES

Shutterstock.com/Nicetoseeya

WORKSHEET 3.1 THE TYPICAL AUSTRALIAN

Page 98

Newspix/Vanessa Hunter

Look at the image depicting the stereotypical Australian and answer the following questions.

consider to think deliberately or carefully about something, typically before making a decision

1 **Consider** who might have created this image and why. What message are they trying to convey?

____

____

propose to put forward (e.g. a point of view, idea, argument, suggestion) for consideration or action

2 **Propose** what the creator of the image wanted the viewer to feel and know.

____

____

____

discuss to talk or write about a topic, taking into account different issues or ideas

3 Do you believe the image portrays 'typical' Australians? Briefly **discuss**.

____

____

____

4 How would you change this image to show 'typical' Australians? It does not need to be a beach scene/backdrop.

____

____

____

____

WORKSHEET 3.2 CLASS SURVEY

SB Page 98

What is the 'average' person in your class like? Construct and conduct a survey of all the students in your class to compile the characteristics of the 'average' student. Your questions will need to be structured to gain the following information:

- age
- gender
- place of birth (and parents' places of birth)
- main language spoken at home
- family structure (e.g. two- or one-parent family, lives with extended family, siblings)
- part-time job and the type of work (e.g. retail, coaching)
- residence (e.g. house, apartment).

You may want to include other questions (e.g. religion, number of cars, number of computers in the household, height).

1 Compile your results and determine the most common response for each question. From the results construct a profile of the characteristics of the 'average' student in your class.

a List the characteristics in the space below.

b Use magazine pictures to make a collage, or draw or make a computer animation of this person.

Characteristics of the 'average' student	Diagram of the 'average' student

WORKSHEET 3.3 DO YOU MEET THE AUSTRALIAN 24-HOUR MOVEMENT GUIDELINES?

Page 107

The Australian 24-Hour Movement Guidelines for Children and Young People recommend that 5- to 17-year-olds include:

- at least 60 minutes of moderate-to-vigorous physical activity every day.
- no more than 2 hours a day screen time for entertainment (e.g. social media, gaming, TV) particularly during daylight hours (unless it is educational).
- An uninterrupted 8 to 10 hours of sleep per night for those aged 14–17 years.

Complete the table below, recalling your physical activity, sedentary behaviour, and sleep for the previous week. Remember to include *incidental* activity (e.g. walking to school, doing chores around the house and garden) as well as *planned* activity (e.g. organised sport and recreation).

	Sunday	Monday	Tuesday	Wednesday	Thursday	Friday	Saturday
Physical activity (PA)	Activities	Activities	Activities	Activities	Activities	Activities	Activities
Time (minutes)							
Total time							
Sedentary behaviour	Activities	Activities	Activities	Activities	Activities	Activities	Activities
Time (minutes)							
Total time							
Start of sleep							
Hours of uninterrupted sleep							

9780170465540

1 Did you meet the Australian 24-Hour Movement Guidelines? Yes/No (circle)

If 'yes', skip Question 2 and answer Question 3. If 'no', go to Question 2.

2 a Why didn't you meet the guidelines? Was it lack of physical activity, too much screen time or not enough sleep?

____

____

b How could you incorporate more physical activity into your day?

____

____

____

3 List three benefits of being physically active. Consider why this might be important to you as an individual, and to your community.

____

____

____

____

____

4 Would you rate your sleep behaviour as positive? Briefly **justify** your answer.

justify
to show how an argument or conclusion is right or reasonable

____

____

____

____

5 Sleep can be interrupted for a variety of reasons. List three reasons contributing to poor sleep and **discuss** what you could do to lessen their influence.

discuss
to talk or write about a topic, taking into account different issues or ideas

____

____

____

____

____

WORKSHEET 3.4 APPS TO MEASURE ACTIVITY

Page 102

Research three different web-based programs or smartphone apps that record, monitor, track or measure physical activity.

1 Find out as much as you can about what each of these programs or apps do, and summarise the information in a Venn diagram to show how they are similar (overlapping areas) and different (open spaces).

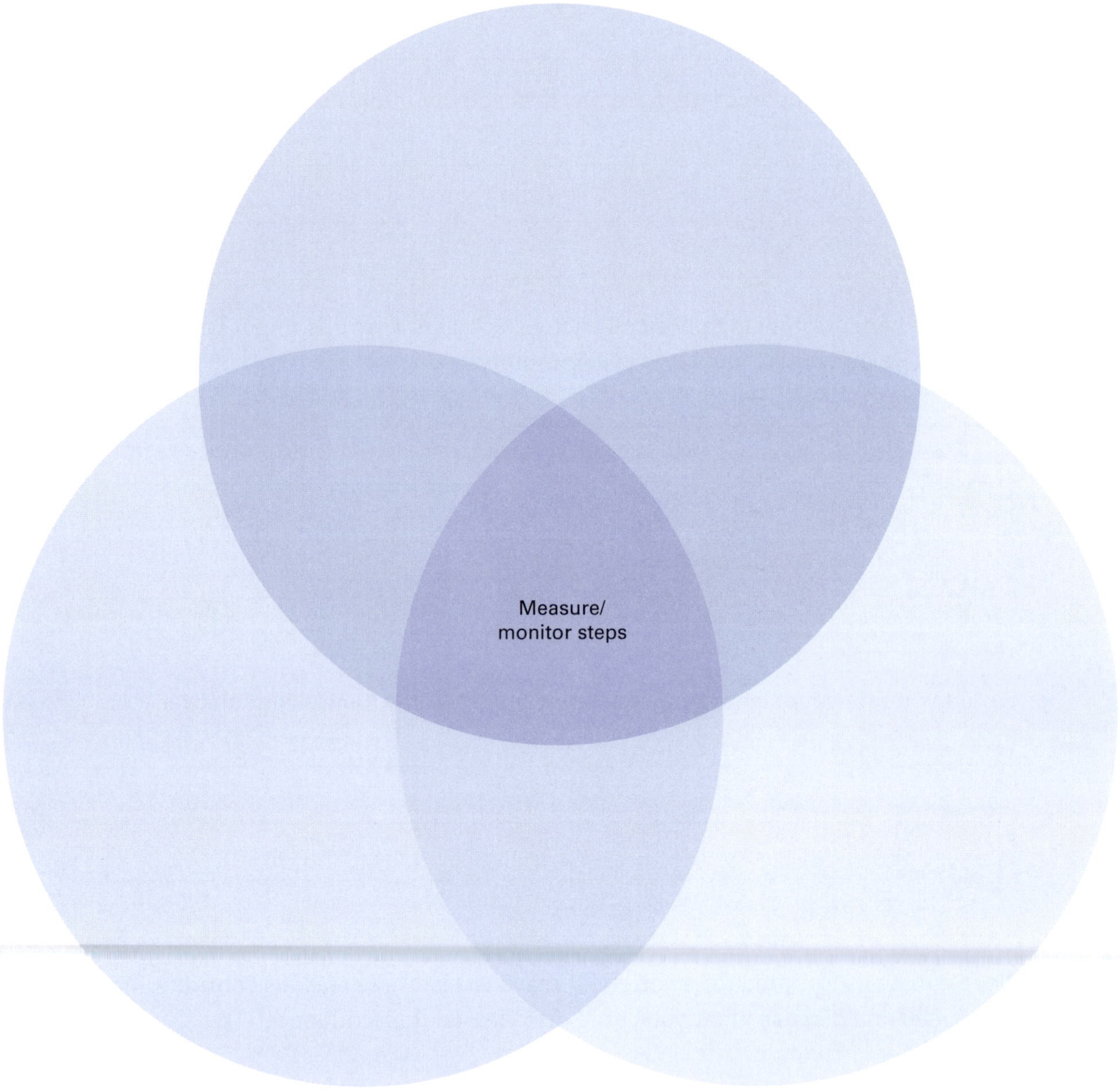

Measure/
monitor steps

2 From your summary above, rank the three programs/apps from least (1) to most (3) useful/user-friendly for a person who is interested in measuring their physical activity levels and **compare** them to the Australian 24-Hour Movement Guidelines for Young People (5–17 years).

compare
to observe or note how things are similar or different

___

___

___

WORKSHEET 3.5 EXTREME TO MAINSTREAM

Page 112

'From extreme to mainstream' is one of the six megatrends thought to be an influence on the future of Australian sport. Many international sporting associations are applying to the International Olympic Committee (IOC) to have their sport included in the Olympic Games. For example, skateboarding and karate were added the Tokyo 2020 program.

Write a proposal to the IOC to have a new sport included in the next Summer Olympic Games (you can choose any sport that is not currently in the Olympics). You will need to research your chosen sport and present a strong argument to the IOC. Your proposal must include a description of the sport, a rationale for its inclusion and a **justification** for why its inclusion in the Olympics will not jeopardise the culture and philosophy of the sport.

justify
to show how an argument or conclusion is right or reasonable

AC

WORKSHEET 3.6 BARRIERS TO BEING ACTIVE

Page 119

What keeps you from being more active?

1 Listed below are reasons that people give to describe why they do not get as much physical activity as they think they should. Read each statement and indicate how likely you are to give these reasons by circling the appropriate number.

Line	Reasons for being inactive	Very likely	Somewhat likely	Somewhat unlikely	Very unlikely
1	My day is so busy now, I just don't think I can make the time to include physical activity in my regular schedule.	3	2	1	0
2	None of my family members or friends like to do anything active, so I don't have a chance to exercise.	3	2	1	0
3	I'm just too tired after school to get any exercise.	3	2	1	0
4	I've been thinking about getting more exercise, but I just can't seem to get started.	3	2	1	0
5	I don't get enough exercise because I have never learnt the skills for any sport.	3	2	1	0
6	I don't have access to jogging trails, swimming pools, bike paths, etc.	3	2	1	0
7	Physical activity takes too much time away from other commitments such as school, family, etc.	3	2	1	0
8	I'm embarrassed about how I will look when I exercise with others.	3	2	1	0
9	I don't get enough sleep as it is. I just couldn't get up early or stay up late to get some exercise.	3	2	1	0
10	It's easier for me to find excuses not to exercise than to go out to do something.	3	2	1	0
11	I really can't see myself learning a new sport at my age.	3	2	1	0
12	It's just too expensive. You have to take a class or join a club or buy the right equipment.	3	2	1	0
13	My free times during the day are too short to include exercise.	3	2	1	0
14	My usual social activities with family or friends do not include physical activity.	3	2	1	0
15	I'm too tired during the week and I need the weekend to catch up on my rest.	3	2	1	0
16	I want to get more exercise, but I just can't seem to make myself stick to anything.	3	2	1	0
17	I'm not good enough at any physical activity to make it fun.	3	2	1	0
18	If we had exercise facilities and showers at school, then I would be more likely to exercise.	3	2	1	0

2 Follow these instructions to score yourself:

a Enter the circled number in the spaces provided, putting together the number for statement 1 on line 1, statement 2 on line 2, and so on.

b Add the three scores on each line. Your barriers to physical activity fall into one or more of six categories: lack of time, social influences, lack of energy, lack of willpower, lack of skill and lack of resources. A score of 5 or above in any category shows that this is an important barrier for you to overcome.

	.....	+	.....	+	.....	=	.....
Line	1		7		13		Lack of time
	.....	+	.....	+	.....	=	.....
Line	2		8		14		Social influence
	.....	+	.....	+	.....	=	.....
Line	3		9		15		Lack of energy
	.....	+	.....	+	.....	=	.....
Line	4		10		16		Lack of willpower
	.....	+	.....	+	.....	=	.....
Line	5		11		17		Lack of skill
	.....	+	.....	+	.....	=	.....
Line	6		12		18		Lack of resources

Source: Adapted from www.health.harvard.edu/newsletters/Harvard_Heart_Letter/2012/January/what-are-your-barriers-to-exercise. Original material from US National Center for Chronic Disease Prevention and Health Promotion's website.

3 Use the website of the US Government Centers for Disease Control and Prevention (search for 'CDC overcoming barriers to physical activity') to find strategies aimed to overcome each of the categories you identified as an important barrier to physical activity for you (the three you scored the highest on). List each of the barriers and the strategy you could implement to overcome this barrier.

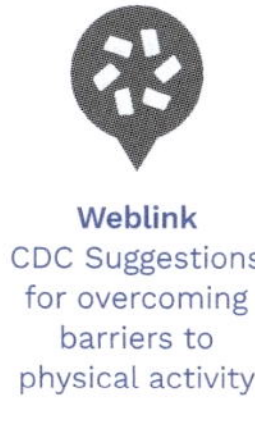

Weblink
CDC Suggestions for overcoming barriers to physical activity

Barrier 1

____

Strategy

____

____

____

Barrier 2

Strategy

Barrier 3

Strategy

4 Extension task

compare to observe or note how things are similar or different

AC

Ask a parent, grandparent or another adult to complete the quiz. **Compare** the perceived barriers of this person with those that you identified as barriers for yourself.

WORKSHEET 3.7 WHO OR WHAT INFLUENCES YOUR PARTICIPATION IN SPORT AND PHYSICAL ACTIVITY?

SB Pages 125–31

1 List the three biggest influences on the sports and physical activities that you do.

2 Categorise these influences as personal, social or cultural.

3 Consider an activity you would like to try. Are there barriers that might prevent you from participating in this activity?

4 Outline how your culture can influence the sport and physical activity you do. Give specific examples from the Australian culture and one other culture you are familiar with.

WORKSHEET 3.7 CONTINUED

5 There are a number of personal factors that influence participation in physical activity, including:

- age
- gender
- socio-economic status
- geographic location
- other personal factors (fundamental motor and sport-specific skills, knowledge, beliefs, abilities/disabilities, motivation).

explain to provide extra information that demonstrates understanding of reasoning and/or application

Explain how the five main personal factors (above) could increase or decrease an individual's involvement in two of the following sports:

- Auskick
- golf
- contemporary dancing
- tennis
- downhill skiing.

Activity 1

____

____

____

____

____

____

____

____

Activity 2

____

____

____

____

____

____

____

____

9780170465540

WORKSHEET 3.8 RETHINKING PHYSICAL EDUCATION AND SPORT

Think about the Physical Education and sport program at your school. Are there ways to make it better? Use the **SCAMPER** (**S**ubstitute, **C**ombine, **A**dapt, **M**odify/**M**agnify/**M**inimise, **P**ut to another use, **E**liminate and **R**emove/**R**everse) framework to think of ways your school could improve PE, sport and physical activity. Write your answers on the following page.

Pages 125–31

		Thinking prompts
S	Substitute (a person or object)	If you were to change one aspect of the program, what would it be and why?
C	Combine (blend and bring together materials, ideas or situations)	Take the best bits of the PE and sport program at your school and adjust them to help increase physical activity levels at your school.
A	Adapt (adjust to new purposes or conditions)	How could you change the PE and sport program to increase physical activity levels?
M	Modify/magnify/minimise (enlarge, reduce or change size, quality or frequency)	What should you do more of, less of, or more often in PE and sport?
P	Put to another use (for another purpose, situation or way of doing things)	How else could the school encourage physical activity, not just in PE and sport?
E	Eliminate (take away or leave out)	What would you leave out from the current PE and sport program? What causes problems in the current program?
R	Remove/reverse (change sequence or layout)	What would happen if PE and sport were removed from the school curriculum?

WORKSHEET 3.8 CONTINUED

S

C

A

M

P

E

R

WORKSHEET 3.9 MEDIA ANALYSIS

Page 133

The *Blueprint for an Active Australia* by the National Heart Foundation of Australia (third edition, 2019) intends to increase levels of physical activity in Australia, leading to community-wide benefits for the environment, social policy and the economy.

The *Blueprint* has 13 key action areas aimed to address barriers to physical activity and to support opportunities for all Australians to be active in their homes, schools, workplaces and communities.

KEY ACTIONS

1. Built environments: Create built environments to support active living.
2. Workplaces: Promote physical activity before, during and after work.
3. Health care: Develop healthcare systems that promote and support physical activity participation.
4. Active travel: Encourage more walking, cycling and public transport use.
5. Prolonged sitting (sedentary behaviour): Promote opportunities and approaches to reduce prolonged sitting.
6. Sport and active recreation: Increase physical activity levels through sport and active recreation.
7. Disadvantaged populations: Address inequality in physical activity participation.
8. Aboriginal and Torres Strait Islander peoples: Provide programs and opportunities to increase physical activity levels among Aboriginal and Torres Strait Islander peoples.
9. Children and young people: Promote healthy development through physical activity participation.
10. Older people: Support healthy and active ageing.
11. Financial measures: Provide financial incentives to make active choices cheaper and easier.
12. Mass-media strategy: Promote the benefits of physical activity.
13. Research and program evaluation: Support the implementation of physical activity initiatives through research, monitoring and evaluation.

Source: *Blueprint for an Active Australia*, 3rd edition, National Heart Foundation of Australia, 2019

For more information, search 'blueprint' on the Heart Foundation website.

Critique the *Blueprint* by answering the following questions.

1 How well do you think the 13 key actions have been implemented?

WORKSHEET 3.9 CONTINUED

2 Identify one mass-media strategy that has been implemented over the past three to five years. List the media avenues that the strategy used to get the message to the wider population. Why is it important that more than one avenue is used?

3 Suggest three ways that the fourth key action has been implemented in your town, city or community.

4 Consider why it is important for First Nations Peoples to be recognised as part of the *Blueprint*.

5 Identify a physical activity program that would be suitable for each of these four age groups: primary school children, teenagers, younger adults, and adults 50 years and older (key actions 9 and 10).

6 Suggest reasons why key settings such as workplaces, schools and childcare facilities are important for physical activity programs.

WORKSHEET 3.10 LADDER OF HEALTH

Rank the following statements from least important (step 1) to most important (step 8) for you to achieve optimal health. Write them beside the steps of the ladder to represent the steps you can take to improve and maintain your health.

Page 134

- Doing at least 60 minutes of physical activity every day
- Eating treats and drinking soft drink in moderation
- Exercising with friends and family
- Being involved in a team sport
- Participating or volunteering in a community-based program
- Having a body mass index (BMI) in the healthy weight range
- Cutting down the time spent on electronic devices for entertainment purposes
- Having 8–9 hours of uninterrupted sleep every night

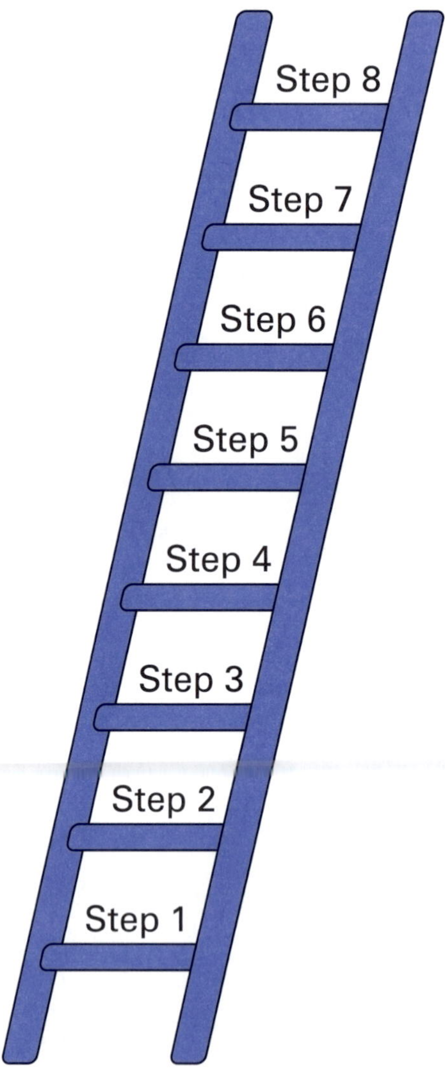

WORKSHEET 3.11 PHYSICAL ACTIVITY – CATERING FOR ALL

Page 134

In the student book you have been encouraged to participate in 'sitting volleyball', which is an adaptive sport that can be played by people with disabilities and enjoyed for health, wellness, leisure, social and competition benefits.

You have also been encouraged to think about ways more young people can engage with sports and may have even participated in modified sports such as Auskick, Netta Netball or Pee Wee Tennis.

In this activity you and a partner are required to modify a sport so it can be more accessible for people with a disability (physical, cognitive, developmental, etc.) and then ask the rest of the class to 'have a go' at your activity.

1 Complete the following table/checklist.

2 Then, with your partner, run the practical activity with your class.

Sport to be modified	
Equipment/space required	
Summary of modification(s)	
Additional rules/ conditions	
Possible challenges in organising this activity for classmates	
Strategies to overcome perceived challenges	

CHAPTER 3 REVIEW

Reflect upon your learning in this chapter by completing the following sentences.

1 My understanding of what is meant by 'healthy people, healthy communities' is ...

2 I now know that ...

3 It was surprising to find out that ...

4 I still have more questions about ...

5 I have changed my opinion about ...

6 The most important thing I have learnt in this chapter is ...

4

DEVELOPMENT AND MANAGEMENT OF MENTAL HEALTH

Shutterstock.com/Nicetoseeya

WORKSHEET 4.1 MYTHS ABOUT MENTAL HEALTH

Page 140

1 Hypothesise why myths are spread about mental health, consider the ones brought up in the 'Bursting the myths' activity and any more you can think of.

2 Propose a reason as to why myths about mental health may increase stigma in the community.

3 Reimagine the myths into a fact that would help prevent stigma in mental health.

9780170465540

WORKSHEET 4.2 GETTING HELP

Choose three of the scenarios below and then complete the table on the following page.

Page 145

SCENARIO 1

Mia has stopped going to dance classes and always makes an excuse not to do PE. She used to love it! You ask Mia what is going on and she tells you that her dance instructor didn't give her a lead role because she couldn't fit into the costume.

SCENARIO 2

Ryan found out that his dad has been diagnosed with a mental illness. Ryan is distraught. He comes to you because he needs some information and doesn't know where to turn.

SCENARIO 3

You haven't seen your friend Sieu-Ting at school for almost two weeks. Her mum calls you one day and asks if you would come over. Your friend confesses to you that every time she starts to get ready for school, she has an anxiety attack and can't move. She just crawls back into bed until it is over. She seems really scared.

SCENARIO 4

Your younger sister tells you that someone is saying hateful things about her online. She is afraid to use the computer because every time she does there's another mean message. You suggest she tells her teacher, but she doesn't want to.

SCENARIO 5

Cooper is devastated because he has just broken up with his partner. No one at school knows that Cooper is gay, except for you. You can't seem to get him interested in anything and are worried about the exams coming up. You know Cooper wants to do well so he can go to university next year to study law.

SCENARIO 6

Faiz has been acting really weirdly; he doesn't spend time with his friends anymore, not even you. You are really worried because he doesn't seem to be looking after himself. He's physically 'there' in class, but his head is somewhere else. The teachers are getting annoyed with him because he just ignores them. When you asked him about it, he got really angry and told you to go away.

WORKSHEET 4.2 CONTINUED

In the table below, recommend the resources that would be good for someone to use in each of your three chosen scenarios. For example, you may recommend that a parent, teacher, doctor, Kids Helpline and the Butterfly Foundation are the best resources someone could use. Label each column with your chosen scenario number and then tick the boxes for the people and resources that apply to each scenario. In the blank space below the table you can write the best website address for the online resources that you recommend. If you think there is another resource that has not been included, write it in the table.

Resource	Scenario	Scenario	Scenario
Parent/carer (specify)			
Friend			
Teacher			
Doctor			
Counsellor: at school or outside			
Lifeline			
Youth Central			
Kids Helpline			
The Butterfly Foundation			
ReachOut			
MindMatters			
Youth Beyond Blue			
headspace			
eSafety Commissioner			
Youth Law Australia			

Useful websites

____

____

____

____

____

WORKSHEET 4.3 TALKIN' ABOUT IDENTITY

Pages 146–7

Choose one of the following songs and find the lyrics online:

- 'Complicated' – Avril Lavigne
- 'Don't let me get me' – P!nk
- 'Born this way' – Lady Gaga
- 'What makes you beautiful – One Direction.

Using one of the song's lyrics:

1 Discuss what the lyrics of this song tell us about identity.

____

____

____

____

____

2 Consider whether these messages are positive or negative and justify why.

____

____

____

____

____

3 Create a list of positive words about identity used in the lyrics.

________	________	________	________
________	________	________	________
________	________	________	________
________	________	________	________

4 Pair up with a peer for a role play. One of you will be someone with poor self-image and the other will be a friend trying to help. Use the list of words above in the role play to try to help your friend feel better about themselves. Circle the words you used in your role play.

5 What other songs are out now that talk about identity in a positive light? Describe what messages they are providing.

____

____

____

WORKSHEET 4.4 IDENTITY

Page 147

Self-esteem, peer pressure, stereotypes, the media and body image all have an impact on your identity to varying degrees. It all depends on how strong your identity is to begin with. If you feel secure about yourself and your identity, then you are less likely to be pressured to do something you don't want to. Read the case study below and then complete the following activity.

CASE STUDY

Sophie is excited to be going to the Year 11 formal; she has a date and has found the most amazing dress. Unfortunately the dress is really expensive and the only size left is a size 8; Sophie is normally a size 10 or 12. The girls she went shopping with all bought their dresses in that store, and they tell Sophie it is easy to get your weight down really quickly. They give her some tips about losing weight.

What should Sophie do?

Identify three options available to Sophie. For each option, come up with three consequences. Finally, make your decision, based on the rest of the table.

Problem: should Sophie buy the dress?	
Options	**Consequences**
	a
	b
	c
Option 2	**a**
	b
	c
Option 3	**a**
	b
	c
Your decision	

WORKSHEET 4.5 SELF-ESTEEM

Page 147

Search online for the video produced by headspace that features Australian teenagers talking about their ideas on self-esteem (see Chapter 4 in the student book). Then answer the following questions.

1 Define self-esteem.

2 Define low self-esteem.

3 Imagine how someone with low self-esteem might feel. Propose some statements they might say to themselves.

4 Consider factors that might lower your self-esteem.

5 Propose three strategies to increase your self-esteem.

6 Investigate three places you could go to for help. Consider both online and in-person resources.

WORKSHEET 4.6 PEER PRESSURE

Page 150

1 In the three scenarios pictured below, a person's friends are pressuring them to have a drink. What might be some risks associated with each choice? Write down the consequences that could occur in each of the choices. Don't forget there will be positive, negative and neutral consequences; don't just think about the negative ones!

Yes, I'll have a drink!

Consequences

____

____

____

____

____

____

No thanks, I don't want one

Consequences

____

____

____

____

____

____

I'm not sure

Consequences

____

____

____

____

____

____

9780170465540

2 How does someone make a decision? Think about how you make decisions; is this the same for everyone? Sometimes you might make different decisions depending on how important the decision is, or how it might affect your life, or for other reasons.

Write down the things that you would consider before making the following decisions. How do you make the final decision?

a Talking to the new student in science class

____

____

____

____

b Going to the local shopping centre after school without your parents' permission

____

____

____

____

c Asking someone out

____

____

____

____

d Drinking alcohol at a party

____

____

____

____

WORKSHEET 4.7 TICKED OFF

Page 152

- ○ What a freak, If I was you I wouldn't show my face
- ○ Where did you learn to do that – in Kindergarten?
- ○ How do you spell ugly? Wait I'm looking at it right now
- ○ Your mumma would be so proud – NOT!!!
- ○ You look like a dirty mess

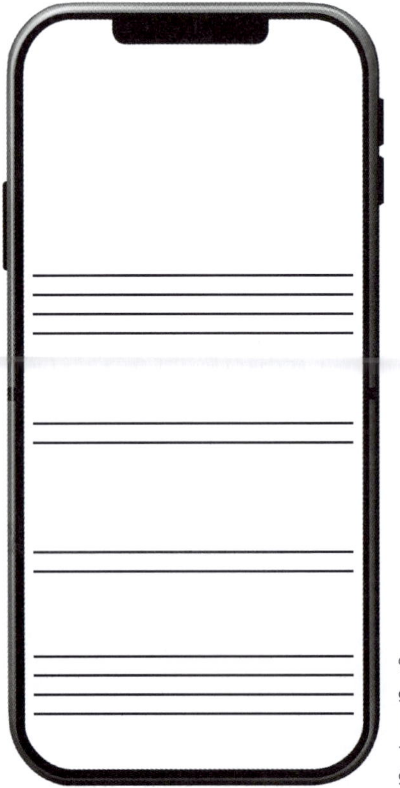

iStock.com/OnstOn

9780170465540

WORKSHEET 4.7 CONTINUED

TikTok is a growing social media platform.

1 Describe the positive aspects of TikTok.

2 Describe the negative aspects of TikTok.

3 Discuss how the comments in Phone 1 would make the creator feel.

4 How do they make you feel without knowing the creator?

5 If you knew that this creator has just spent five days learning a dance to put up on TikTok. It was the first time that they had the confidence to post and thought that it was pretty good! They had shown it to a trusted friend who also said it looked good and yes, they should go ahead and post it. You may or may not like the post/dance however, you have noticed there has been some nasty, even bullying comments about this post. You decide to write some positive comments to help boost their self-esteem. Write these comments on Phone 2.

6 What would you do if you saw negative comments like this on your own posts?

WORKSHEET 4.8 NEGATIVE BODY IMAGE

Pages 58–60

Australia is a diverse nation and includes people:

- from many cultures
- with varying physical characteristics
- of varying ages
- with varying educational attainments
- with varying abilities
- with varying skills
- with varying resources
- with varying morals/values/standards.

1 **a** Consider whether advertising celebrates diversity. (Does it depict the wide range of people we have in our country?)

Yes/No (circle one)

b Justify your thinking here:

____

____

____

____

2 **a** Look online, in magazines or on TV to see how advertising celebrates diversity. Do these examples show diversity in advertising?

Yes/No (circle one)

describe
to give an account of characteristics or features

b Cut out and attach, or **describe** some examples.

____

____

____

____

____

____

____

____

9780170465540

3 Write a paragraph defending one the following statements. (Circle the statement that you are using.)

a Advertising celebrates diversity.

b Advertising does not celebrate diversity.

4 Discuss whether there are any examples that show the other side of the story. Describe how.

WORKSHEET 4.9 REDUCING STRESS DURING EXAM TIME

Pages 161–4

Exam time can be very stressful for some people, even if they feel they have prepared enough. Stress is a normal bodily response that can present differently in each person. Some people are also good at covering up their stress levels; others may not be able to pretend they are fine.

Imagine you are just about to take an important exam that you want to do well in. On the continuum below, mark an appropriate point for how stressed you would be.

Relaxed, not feeling stressed at all ←――――――――――→ Stressed out to the max

1 What signs of stress do you usually feel? For example: heart racing, being forgetful, tearing up or feeling sleepy.

____

____

____

____

2 Identify four things you can do in the weeks leading up to an exam that will help you feel less stressed.

____

____

____

____

3 Identify three things that you can do the evening before the exam that will help you feel less stressed.

____

____

____

4 Identify two things you can do on the morning of the exam that will help you feel less stressed.

____

____

5 Identify one thing you can do when you begin the exam to help you feel less stressed.

____

9780170465540

WORKSHEET 4.10 PROMOTING HEALTH

Page 172

The World Health Organization identified 10 key action areas for health promotion.

- Build healthy public policy
- Create supportive environments
- Strengthen community action
- Develop personal skills
- Reorient health services towards primary health care
- Promote social responsibility for health
- Increase investments for health development to address social inequities leading to poor health
- Consolidate and expand partnerships for health
- Strengthen communities and increase community capacity to empower the individual
- Secure an infrastructure for health promotion

Choose one of these action areas and write about why the action area is important to address to improve mental health. For further detail, google 'Jakarta declaration on leading health promotion into the 21st century' to find the page on the World Health Organization's website.

WORKSHEET 4.11 HEALTH BEHAVIOURS INFLUENCING MENTAL HEALTH

Pages
173–5

Weblink
Extension: Read more about this research on health behaviours and mental health

A study in New Zealand in 2020 looked at which health behaviours (sleep, physical activity and diet) had the greatest impact on young people's mental health and wellbeing. The study found that sleep was the most important factor. In fact, sleep was broken into two areas, sleep quality and sleep quantity.

The study concluded:

Sleep quality is an important predictor of mental health and well-being in young adults, whereas physical activity and diet are secondary but still significant factors. Although strictly correlational, these patterns suggest that future interventions could prioritize sleep quality to maximise mental health and wellbeing in young adults.

Source: Wickham S-R, Amarasekara NA, Bartonicek A and Conner TS (2020). The big three health behaviors and mental health and well-being among young adults: A cross-sectional investigation of sleep, exercise, and diet. *Front. Psychol.* 11:579205. doi: 10.3389/fpsyg.2020.579205

1 Describe your sleeping habits in terms of sleep quality and sleep quantity.

a Quality

____

____

b Quantity

____

____

2 Describe any factors that impact the quality or quantity of your sleep. Consider any factors that may occur day-to-day, occasionally or rarely.

____

____

____

____

____

3 Create a list of your 'Top 3 tips for getting a good night's sleep'.

____

____

____

____

____

9780170465540

WORKSHEET 4.12 BROCHURE EVALUATION

4

SB
Page 178

Weblink
View the Yarning about sadness brochure online

Look at the following two brochures and then answer the questions below.

BROCHURE 1:

Yarning about sadness by the Menzies School of Health Research is an Aboriginal and Torres Strait Islander mental health initiative.

Yarning about sadness

Depression or sadness

Our people have strong culture. We are artists and storytellers, we are sporting legends and skilled hunters, we are musicians and dancers and uncles and aunties and grandmothers and grandfathers.

Most of all we are teachers, and we are teaching our children to find their way in a modern world. Our kids need a guide to find their way in the modern world ... they need to take our culture with them ... to bring both worlds into one.

What is mental health?

Mental health and well-being is like a tree with four branches which needs to be looked after. When the tree is well balanced, a person's mental health is strong. What's around us, what we do, what we think and what we feel helps to keep us strong.

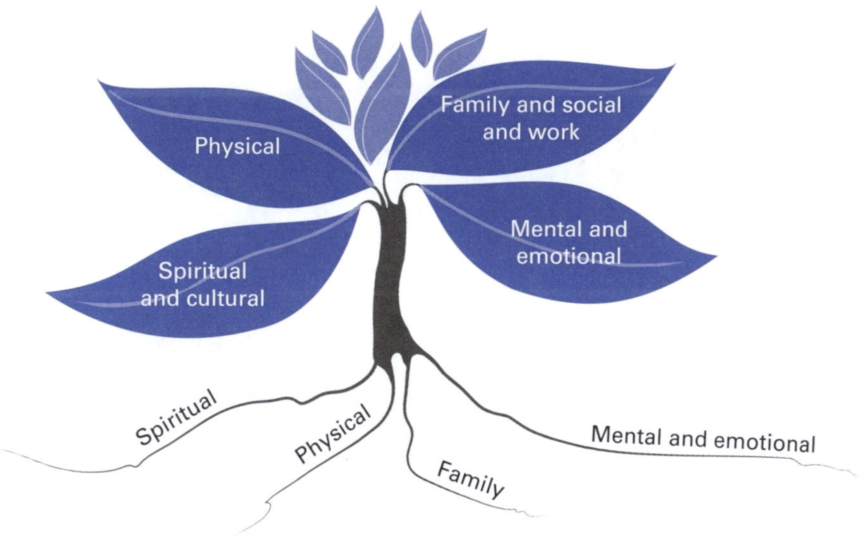

What causes depression or sadness?

We all have things that take our strength away. Worries from now and worries from the past. Too many worries and troubles can take balance away. When we are out of balance we can get depression or sadness.

Aboriginal and Torres Strait Islander people can feel out of balance when modern life takes them away from what makes them strong:

- Country
- Family
- Culture
- Language

These things can cause depression and sadness:

- Poor physical health
- Loss or bereavement
- Stress
- Too much alcohol or ganja (marijuana) or other drugs
- Family history of mental illness
- Stopping usual treatment for mental illness

Getting out of balance

People with depression or sadness can think differently, feel differently and behave differently.

Sad thoughts might be:

- I can't do anything right
- It's all my fault
- No-one cares
- Things will never be better

Sad feelings might be:

- Feeling nervous and worrying all the time
- Feeling guilty and worthless
- Crying and feeling sad

When we have too many worries we may get out of balance:

- Can't sleep
- Change of appetite (eat more or less than usual)
- Lose interest (nothing is fun)
- Cry for no reason
- Sit down alone
- Negative thoughts and feelings

If we don't talk to someone it can lead to other problems:

- Violence
- Self harm

- Trouble parenting
- Family worry
- Substance misuse

Worries that can take away our strength:

- culture worry, not hunting, not exercising, not taking medication, no good tucker, physical illness, memory worry;
- sleep worry, ganja (marijuana), grog, being alone, side effects, work worry, too much energy, gambling, family worry, don't know enough about illness;
- anxious, violent, not caring for self, sad, mixed up thoughts, hearing voices, self-harm

We often cover up how we are feeling, and sometimes it's hard for others to know there is anything wrong.

There are four things to do to treat depression and to get back in balance again:

- Talk to someone – family or friends
- Do more things that keep you strong
- Do less of the things that take your strength away ... and if that's not working
- Try talking with a health professional

Growing strong again

There can be many treatments to help people to grow their spirit back to strength

- Education can help change how a person feels and help them make good choices that make their spirit stronger.
- Medications can help improve the symptoms of depression and sadness so people can focus on growing stronger
- Knowing early warning signs of stress can help us be prepared

Where to get help or information

Talk to someone you trust: a friend, family member, senior elder, traditional healer or visit your local Aboriginal and/or Torres Strait Islander Health Service or local health centre.

Lifeline 131114 or www.lifeline.org.au
Beyondblue 1300224636 or www.beyondblue.org.au
Tamarind Centre NT 89994988
Top End Mental Health Services 08 8999 4988
www.menzies.edu.au/AIMHI

Source: Menzies School of Health Research, 'Yarning about sadness' brochure, www.menzies.edu.au/icms_docs/161349_Yarning_about_sadness_brochure.pdf

BROCHURE 2:

A fact sheet from the National Institute of Mental Health.

Depression and high school students: Answers to students' frequently asked questions about depression.

Depression can occur during adolescence, a time of great personal change. You may be facing changes in where you go to school, your friends, your after-school activities, as well as in relationships with your family members. You may have different feelings about the type of person you want to be, your future plans, and may be making decisions for the first time in your life.

Many students don't know where to go for mental health treatment or believe that treatment won't help. Others don't get help because they think depression symptoms are just part of the typical stresses of school or being a teen. Some students worry what other people will think if they seek mental health care.

What is depression?

Depression is a common but serious mental illness typically marked by sad or anxious feelings. Most students occasionally feel sad or anxious, but these emotions usually pass quickly – within a couple of days. Untreated depression lasts for a long time and interferes with your day-to-day activities.

What are the symptoms of depression?

Different people experience different symptoms of depression. If you are depressed, you may feel:

- Sad
- Anxious
- Empty
- Hopeless
- Guilty
- Worthless
- Helpless
- Irritable
- Restless

You may also experience one or more of the following symptoms:

- Loss of interest in activities you used to enjoy
- Lack of energy
- Problems concentrating, remembering information or making decisions
- Problems falling asleep, staying asleep or sleeping too much
- Loss of appetite or eating too much
- Thoughts of suicide or suicide attempts
- Aches, pains, headaches, cramps or digestive problems that do not go away

Depression in adolescence frequently co-occurs with other disorders such as anxiety, disruptive behaviour, eating disorders or substance abuse. It can also lead to increased risk for suicide.

Are there different types of depression?

Yes. The most common depressive disorders are:

- **Major depressive disorder** – also called major depression. The symptoms of major depression are disabling and interfere with everyday activities such as studying, eating, and sleeping. People with this disorder may have only one episode of major depression in their lifetimes. But more often, depression comes back repeatedly.
- **Dysthymic disorder** – also called dysthymia. Dysthymia is mild, chronic depression. The symptoms of dysthymia last for a long time – 2 years or more. Dysthymia is less severe than major depression, but it can still interfere with everyday activities. People with dysthymia may also experience one or more episodes of major depression during their lifetimes.
- **Minor depression** – similar to major depression and dysthymia. Symptoms of minor depression are less severe and/or are usually shorter term. Without treatment, however, people with minor depression are at high risk for developing major depressive disorder.

United States National Institute of Mental Health

1 Which brochure do you like best? Why?

____

____

____

____

2 What are the differences between the brochures? Why do you think these differences exist?

____

____

____

____

____

____

____

____

____

3 As a mental health promotion document, what do you think each brochure does well?

4 As a mental health promotion document, what do you think each brochure doesn't do well (if anything)?

5 Which brochure do you think would be most useful for your school? Why?

WORKSHEET 4.13 MENTAL HEALTH DAY

Page 178

World Mental Health Day (WMHD) occurs every year on 10 October. Each year there is a new campaign focusing on one aspect of mental health. For example, suicide prevention, young people and mental health, and challenging perceptions about mental illness.

Find out more about the current campaign in Australia by going to 1010.org.au.

You can also google 'World Health Organization World Mental Health Day'. Then, answer the questions below.

1 The principal asks you to be in charge of organising a Mental Health Day at your school. Using what you have learnt so far, what would your goals be for your school?

a Consider what information you would need to gather to help with planning the event.

____

____

____

____

b List what your objectives would be.

____

____

____

____

c Discuss what your theme would be.

____

____

d Outline your target audience

____

____

e Discuss how you would get people to help and participate.

____

____

____

____

____

WORKSHEET 4.13 CONTINUED

f List the types of activities/classes/resources you would provide.

g Develop a plan of how you would promote your event.

h Discuss where you would hold your event, and who would you invite.

i Determine where the event would take place.

j Create a program of events. What would be happening during the event?

k Discuss how you will evaluate it (how will you know if it was successful).

2 Is there anything else you would need to consider?

3 Research what the focus for WMHD is this year and summarise it below.

CHAPTER 4 REVIEW

Reflect upon your learning in this chapter by completing the following questions.

1 Think about four people in your life. They could be friends, family or others. How do their identities have an impact on your own identity?

2 Young people are bombarded with gender stereotypical messages in the media. Find an advertisement that shows gender stereotyping and answer the following questions:

a Why do you think this advertisement is showing gender stereotyping?

b What messages is it sending out?

c How do you think this image might affect someone's identity?

3 If a friend came to you for support or help, what would be the first three things you would do?

4 What is empathy and how does it affect relationships?

5 What are the first three community resources you would research to get help with a mental health issue?

6 Identify five new activities your school could implement to help reduce the stigma of mental health or to highlight the need for getting help early for mental health issues.

7 List the three most important things you have learnt from this chapter.

8 Is there anything else you need to know about mental health? Do you have any questions about mental health? Write them here and don't forget to ask your teacher.

9780170465540

5

SEXUALITY, GENDER AND RESPECTFUL RELATIONSHIPS

Shutterstock.com/Nicetoseeya

WORKSHEET 5.1 INFLUENCES ON SEXUALITY

Page 187

describe
to give an account of characteristics or features

Imagine you are able to plan a ceremony for your 'coming of age' (your transition from a child to adult). How would you like to celebrate this milestone? Would there be any expectations from your family, friends, culture, religion or school that you may need to consider? **Describe** or draw your 'coming of age' ceremony.

WORKSHEET 5.2 SEXUAL ORIENTATION

The number of same-sex couples declared in the census in Australia has been on the rise since 1996.

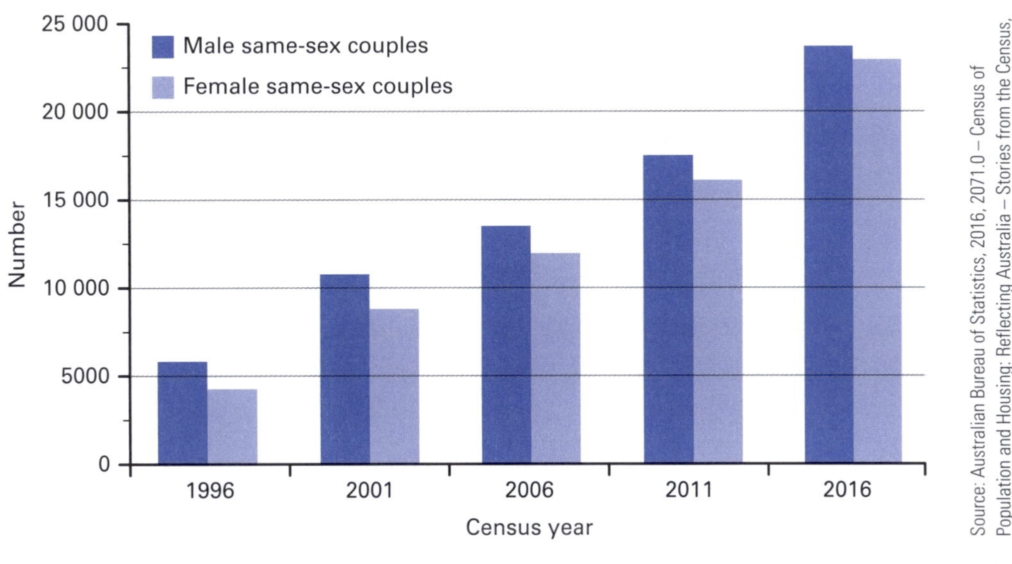

Source: Australian Bureau of Statistics, 2016, 2071.0 – Census of Population and Housing: Reflecting Australia – Stories from the Census, 2016, viewed 28 April 2020, https://www.abs.gov.au/ausstats/abs@.nsf/Lookup/by%20Subject/2071.0~2016~Main%20Features~Same-Sex%20Couples~85

1 Do you think there are more same-sex couples than ever before or do you think that it is more acceptable to publicly identify as being in a same-sex relationship? Each balloon represents one option. Brainstorm reasons why people might argue for each side.

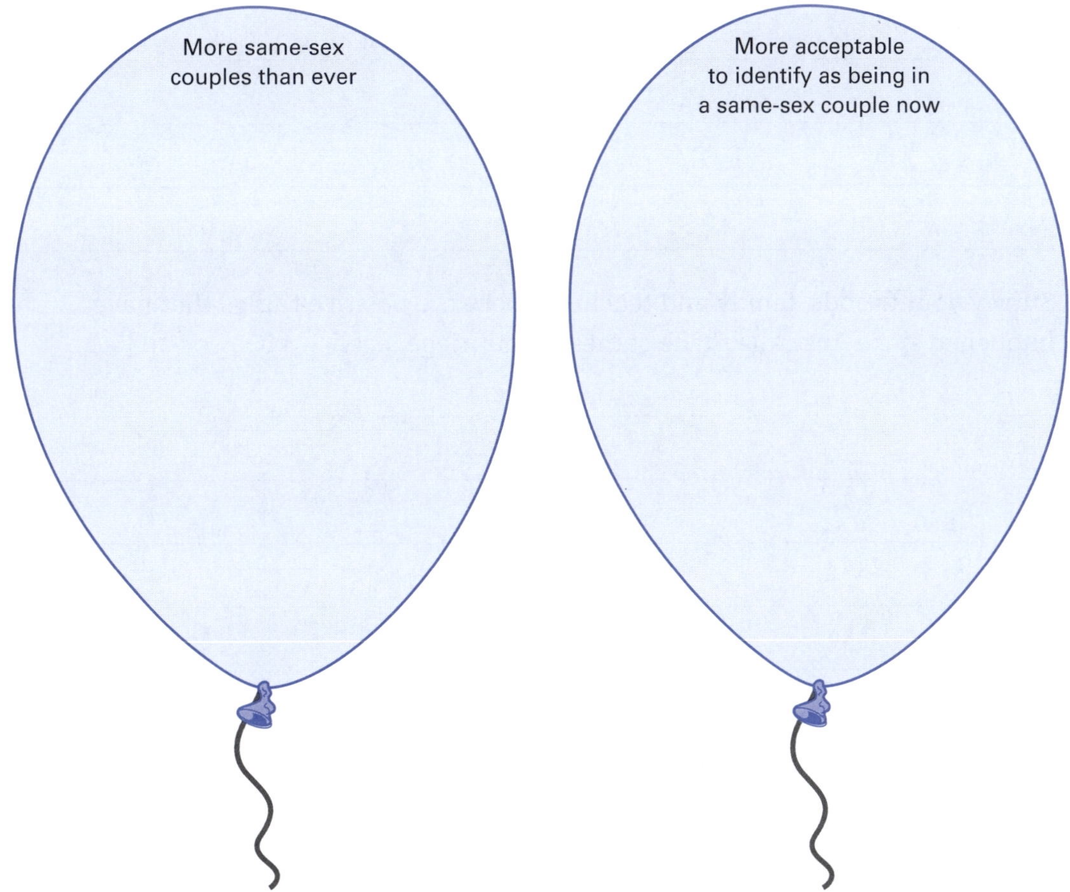

WORKSHEET 5.2 CONTINUED

2 Consider how Australia's same-sex marriage laws will have an impact on the number of same-sex couples reported at the next census.

3 Same-sex marriage is still illegal in some countries. Make a list of reasons why this might be.

4 Survey your friends, family and teachers about the positive things that have happened since Australia made same-sex marriage legal.

WORKSHEET 5.3 STEREOTYPES

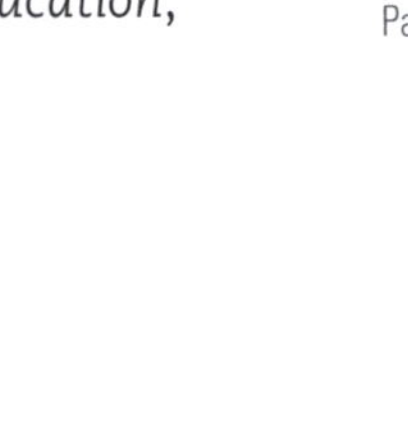

SB Page 189

1 Think of the popularity of celebrities who have found fame on *RuPaul's Drag Race*, including Australian Courtney Act. Then there are actors such as Billy Porter and Cara Delevingne or characters in popular TV shows, such as Eric in *Sex Education*, who all challenge gender stereotypes.

Individuals outside traditional gender stereotypes, clockwise from top left: Courtney Act, an Australian performer who shot to fame on *RuPaul's Drag Race*; Billy Porter, stage and screen actor; the character Eric Effiong (played by Ncuti Gatwa) from popular Netflix show *Sex Education*; Cara Delevingne, openly genderfluid actor and model.

a Consider why these individuals are so popular now.

____

____

____

____

____

____

____

b Discuss whether you think the popularity of gender-diverse celebrities might impact future gender stereotyping in the media.

2 Consider all the media you are interested in (YouTube, social media, movies, TV shows, websites and magazines) as well as advertising displayed on these platforms. Discuss whether the articles, shows or advertising conform to gender stereotypes. Can you find any examples where someone is shown as being or acting 'outside the norm'? **Describe** the advertisement, article or show (or cut it out and attach it from a magazine) and **explain** why a non-stereotypical representation of gender is used in this case.

describe to give an account of characteristics or features

explain to provide extra information that demonstrates understanding of reasoning and/or application

AC

9780170465540

WORKSHEET 5.4 ADOLESCENCE

SB Page 192

As adolescents mature, they become more independent.

1 List all the ways that you will become independent over the next five years.

In one year

In two years

In three years

In four years

In five years

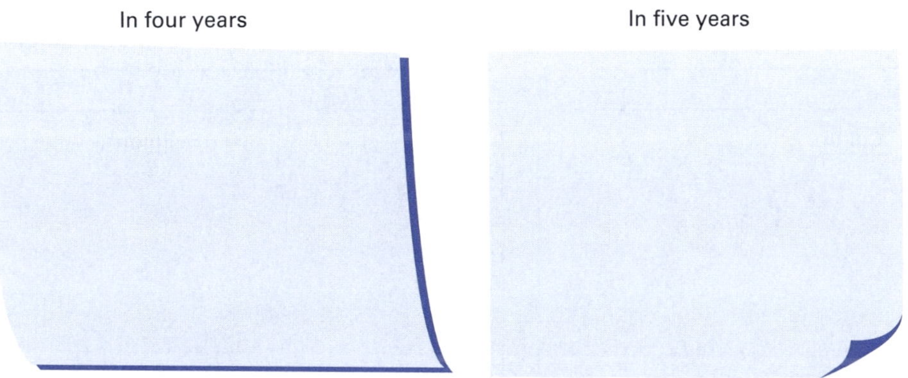

And with independence comes responsibility.

2 Consider all the physical, social and emotional ways in which you have or will become more independent. Fill in the table below; an example has been given.

3 Think about the responsibility that is attached to that independence. Add to the table below; an example has been given.

	Independence	Responsibility
Social	*Going to the movies with my friends*	*Behaving in an appropriate manner at the movies and waiting for parent/carer to pick me up where we agreed*
Physical		
Social		
Emotional		

WORKSHEET 5.5 PERSONAL IDENTITY

Page 195

1 Are there expectations on you to do or be anything? For example, are you expected to behave in a certain way, attend certain events, or excel at things? In the corresponding box, list some expectations of you from the group identified. As an example, your teachers expect you to be at school! Using the last box, identify another group and its expectations of you.

Family	Friends

Culture	Religion

Teachers/ school	Other

2 Consider why people expect things of you. Choose one example from each group and **explain** why you think they have those expectations.

explain to provide extra information that demonstrates understanding of reasoning and/or application

a Family

____

____

____

b Friends

c Culture

d Religion

e Teachers/school

f Your other group

3 a Justify whether realistic expectations help us. Do they motivate us to do better or try new things?

b Consider if there were no expectations on you – how might that affect your life?

WORKSHEET 5.6 FORMING RELATIONSHIPS

Page 199

Starting a relationship is a very exciting time. However, it can be confusing unless you are very clear with your partner about what is going on.

1 How do you know if you are ready for a relationship?

2 How do you know if you are in a relationship?

3 What types of feelings do you experience or do you expect to experience when you start a relationship?

4 How do you ensure that you still spend time with your friends and family when you are in a relationship?

5 Do you think that young people might give in to peer pressure to start dating or participating in sexual activity because everyone else is? Why or why not?

6 What would you say to a friend who is being pressured into starting a sexual relationship when you know they aren't ready yet?

9780170465540

WORKSHEET 5.7 GENDER STEREOTYPES

Page 205

1 In each of the boxes below, write as many things that you can think of that portray a stereotypical 'man' (e.g. strong) or a stereotypical 'woman' (e.g. weak). Consider physical, emotional, sexual, financial, behavioural, educational and employment stereotypes.

2 Sometimes people who 'step outside the box' will get called names because they aren't seen as 'conforming'. Why do people call other people names? Why do people pressure others to stay in the box?

____

____

____

____

3 Underneath each box, list at least 10 things that a man or woman can be or do that are not stereotypical. This could be a characteristic (e.g. a man can be emotional), an interest (e.g. a woman can like beer) or a career (e.g. a woman can be a CEO).

4 Is there anything wrong with the people who aren't in the box?

____

____

5 If you could choose someone to be a friend or intimate partner with, would you choose characteristics from both inside and outside the box? **Explain** why.

____

____

____

explain
to provide extra information that demonstrates understanding of reasoning and/or application

AC

Stereotypical man	Stereotypical woman

WORKSHEET 5.8 ADDRESSING THE DRIVERS OF GENDER-BASED VIOLENCE

Page 206

THE PROBLEM

Violence against women is serious, prevalent and driven by gender inequality

Gendered drivers of violence against women

Driver 1.	Driver 2.	Driver 3.	Driver 4.
Condoning of violence against women	Men's control of decision-making and limits to women's independence in public and private life	Rigid gender stereotyping and dominant forms of masculinity	Male peer relations and cultures of masculinity that emphasise aggression, dominance and control

Underlying **social context** for violence against women

SOCIAL CONTEXT

Gender inequality and other forms of **oppression**

such as racism, ableism, ageism, classism, cissexism and heteronormativity

Factors that reinforce violence against women

REINFORCING Factor 1.	REINFORCING Factor 2.	REINFORCING Factor 3.	REINFORCING Factor 4.
Condoning of violence in general	Experience of, and exposure to, violence	Factors that weaken prosocial behaviour	Resistance and backlash to prevention and gender equality efforts

These drivers and reinforcing factors play out at every level of society: from individual attitudes and behaviours, to social norms, organisational cultures and practices, policies, laws, and institutions.

9780170465540

THE SOLUTION

Violence against women is preventable if we all work together

Essential actions to address the gendered drivers

ACTION 1. ESSENTIAL
Challenge the condoning of violence against women

ACTION 2. ESSENTIAL
Promote women's independence and decision-making in public life and relationships

ACTION 3. ESSENTIAL
Build new social norms that foster personal identities not constrained by rigid gender stereotypes

ACTION 4. ESSENTIAL
Support men and boys in developing healthy masculinities and positive, supportive male peer relationships

Essential actions to address the underlying social context

ACTION 5. ESSENTIAL
Promote and normalise gender equality in public and private life

ACTION 6. ESSENTIAL
Address the intersections between gender inequality and other forms of systemic and structural oppression and discrimination, and promote broader social justice

ACTION 7. ESSENTIAL
Build safe, fair and equitable organisations and institutions by focusing on policy and systems change

ACTION 8. ESSENTIAL
Strengthen positive, equal and respectful relations between and among women and men, girls and boys, in public and private spheres

Supporting actions to address the reinforcing factors

ACTION 9. SUPPORTING
Challenge the normalisation of violence and aggression as an expression of masculinity

ACTION 10. SUPPORTING
Reduce the long-term impacts of exposure to violence, and prevent further exposure

ACTION 11. SUPPORTING
Strengthen prosocial behaviour

ACTION 12. SUPPORTING
Plan for and actively address backlash and resistance

These 12 actions need to be implemented at every level of society: using legislative, institutional, policy and program responses; by governments, organisations and individuals; in settings where people live, work, learn and socialise; in ways that are tailored to the context and needs of different groups.

Source: Our Watch. (2021). Change the story: A shared framework for the primary prevention of violence against women in Australia (2nd ed.). Melbourne, Australia: Our Watch., https://www.ourwatch.org.au/change-the-story/

PREVENTING VIOLENCE AGAINST WOMEN

explain
to provide extra information that demonstrates understanding of reasoning and/or application

1 Have a look at the drivers of violence against women in the table. Can you think of an example for each of the drivers? Think about the different areas in which women and men work, play, live and love; for example, sports, the law, politics, education, cultural or religious institutions, the home and childcare.

a ____________________

____

b ____________________

____

c ____________________

____

d ____________________

____

2 Discuss whether you have seen anything in the media that tries to address the drivers and prevent violence against women.

____

____

3 Search YouTube for the five 'Stop it at the Start' TV adverts created by the Australian Government to help prevent gender-based violence. For each video **explain** what is happening and why the behaviour, and how it is dismissed, condones violence against women.

a ____________________

____

____

b ____________________

____

____

c ____________________

____

____

 9780170465540

d ____

____

____

e ____

____

____

4 Make a list of examples of any actions in your day-to-day life that help prevent violence against women.

____

____

____

5 Propose any examples of actions that could be done in your day-to-day life that would help prevent violence against women.

____

____

____

____

____

WORKSHEET 5.9 HARASSMENT

Page 206

Identify whether harassment occurs in the five scenarios described below. If it is harassment, recommend what the victim should or could do to improve their situation If it is not harassment, explain ways that the individual could explain this to someone.

SCENARIO 1

Tran goes into work every day dreading what might happen. His female boss makes inappropriate comments about what she does on the weekend with her partners. She is very vocal about it and others in the office laugh it off. Tran thinks she knows that he finds it awkward, because he doesn't say anything, so she has been coming into his office to ask him what he gets up to on the weekend. Tran doesn't know what to do. Everyone else seems to think it is fine – should he just ignore her?

____

____

____

____

SCENARIO 2

Emma has been feeling really down at school. Almost every day she hears someone say 'That's so gay!' about something that they think is bad, awful or stupid. Emma is gay, although no one else knows yet. So, when they say that, it makes her feel as if she is bad, awful or stupid.

____

____

____

____

SCENARIO 3

Eli is really smart; he knows exactly what he wants to do in life. He has it all mapped out! The other kids are amazed by his drive and determination. They call him 'Bookie' because he always has his head in a book and won't muck around with them. Eli wasn't sure about the name at first, but has realised that the kids actually like him and gave him a nickname to help him fit in; it was the first one he'd ever been given. Eli's mum asks him if he finds the nickname offensive. Eli says, 'No, I actually like it'.

9780170465540

SCENARIO 4

Alf gets called 'Ranga' because of his red hair. Everybody really likes Alf and they call him that affectionately. Alf can't stand it; it makes him angry every time someone uses the name. However, he wears glasses also, and thinks 'Ranga' is probably better than being called 'four-eyes'.

SCENARIO 5

Every time Kate goes to her locker and passes Ben in the corridor, Ben reaches out and pinches her bottom. Her friends think she is really lucky because Ben must like her; he's the hottest boy in school. Kate does not like it at all.

SCENARIO 6

Jessie plays netball every Saturday at the local courts. A group of guys from another school come every weekend to watch her play. Every time she gets the ball they whistle and cheer loudly. They sit down the far end away from the other spectators. Jessie doesn't know why they come to watch her and feels weird when everyone asks who they are.

WORKSHEET 5.10 MAKING DECISIONS

Page 210

Your friend Sam has been going out with Alex for six months. They have not gone beyond kissing and cuddling. Alex is urging Sam to start a sexual relationship, and Sam feels Alex will break up the relationship if it doesn't progress in this way.

describe to give an account of characteristics or features

1 **Describe** the decision Sam has to make.

2 Make a list of the options Sam has. Evaluate the risks and consequences of each option.

3 Describe two long-term impacts if Sam decides to start a sexual relationship with Alex.

4 If Sam and Alex were of the opposite sex, would the decision-making process be different? Evaluate why or why not.

Sam comes to you and says, 'I am thinking about having sex, but I'm not sure if I am totally ready.'

5 What four questions would you ask Sam that may help them make a decision?

a

b

c

d

WORKSHEET 5.10 CONTINUED

6 Brainstorm the sorts of things you would expect Sam to say if Sam was totally ready.

7 As with most activities, a range of behaviours can be associated with sexual activity. These range from abstinence to risky sexual practices. Justify why people choose:

a to practise abstinence?

b to carry out safer sex practices?

c to engage in risky sexual practices?

8 List four things people can do to make any sexual experience safer (think about physical, social, emotional and mental safety).

a

b

c

d

WORKSHEET 5.11 LAWS OF CONSENT

1 Define 'age of consent'.

Weblink
Access the following website to see the age of consent by country

2 Scroll down to Australia to see the laws that apply here.

3 Read the following countries' descriptions and make notes about the differences in age of consent to Australia and any other interesting facts:

- South Korea
- Pakistan
- Philippines
- India

4 Consider what factors have an influence on the different age of consent laws around the world.

9780170465540

5 Access the following article about changes in the law to affirmative consent. Discuss how you think this might change the story in Australia.

Weblink
Affirmative consent law reforms

WORKSHEET 5.12 CONSENT

Page 210

When individuals start a sexual relationship, they *must* get consent before participating in sexual activity; otherwise, it is against the law.

1 What is the definition of consent?

______

______

______

2 Apart from being the law, why do you think it is so important to get consent before you participate in sexual activity?

______

______

______

______

______

3 How do you get consent?

______

______

______

4 What happens if someone doesn't say anything, even when you have asked them a few times?

______

______

______

5 a If someone gives consent is it OK for them to change their mind and say 'no' later?

Yes/No (circle one)

b Why might someone change their mind?

______

______

______

9780170465540

6 Sometimes a person cannot give consent and therefore no sexual activity should occur. List all the examples you can think of where someone cannot give consent.

____

____

____

____

____

7 Write down as many 'pressure lines' that you can think of (these are lines that people will use to try to convince you to do something). For example, 'Everyone else is doing it, why can't we?' For each pressure line come up with a response.

Pressure lines	Response to pressure lines

8 If you heard a friend saying these pressure lines to their partner, what would you say to them?

____

____

____

9 Discuss why you think some people find it hard to say what they want. What do you think they are worried about?

____

____

____

10 Create a 2-minute rap/song, speech, poem or play about how people must obtain consent before engaging in sexual activity with each other. Do not assume silence is consent. When giving consent each person must say 'Yes'. If someone doesn't say 'Yes' out loud there is no consent.

9780170465540

WORKSHEET 5.13 CONSENT IN PRACTICE

SB Page 210

Read the scenario below and answer the following questions.

SCENARIO

Kit went to a party last night. It was fun until Kit had too much to drink. Kit and Hossein went upstairs to have sex. Neither Kit nor Hossein planned on doing this, so neither had brought protection. Hossein tried to tell Kit that they should stop, but it looked like Kit was really enjoying it, so Hossein shut up. The next day both Kit and Hossein felt really bad about what had happened, but were too embarrassed to call each other. They were really surprised when they looked at their Facebook page and one of their 'friends' had posted 'Who went upstairs for some private FUN?' and tagged them both.

1 If you were Kit or Hossein, what would you do now?

2 What impact did alcohol have on this scenario?

3 Is this unwanted sex? What does the law say about consent?

4 Are you allowed to post something about someone else without getting their permission?

WORKSHEET 5.13 CONTINUED

5 Consider the risks present in this scenario.

6 Outline two considerations partners should discuss with each other before sexual activity occurs.

7 Hypothesise how this situation could have been handled differently.

8 After this incident, do you think Kit and Hossein will begin a relationship or stay together if they are already in one? Why or why not?

CHAPTER 5 REVIEW

Reflect upon your learning in this chapter by completing the following questions.

1 List at least five important qualities of a relationship. Then rank these qualities from MUST HAVE in a relationship to WOULD LIKE in a relationship.

Qualities	Order of necessity	
	MUST HAVE	WOULD LIKE

2 List four things that you have learnt from this chapter.

---

3 Is there anything you will do differently because of what you have learnt in this chapter?

---

4 List the five most important things you think young people should know about relationships and sexuality.

5 Write or draw some potential benefits of being in an intimate relationship (consider physical, mental, emotional and social benefits).

6 Write or draw some potential risks of being in an intimate relationship (consider physical, mental, emotional and social risks).

7 If you find yourself in a situation where you feel you are being harassed, what could you do, and where could you go?

9780170465540

8 Consider why getting consent for sexual activity is so important. List five benefits of asking for consent.

9 There are good resources that provide information about relationships and sexuality. However, this topic can be sensitive or complicated and may require more than just general information.

a What signs would tell you that a resource is reliable or trustworthy?

b If you don't think a resource is giving you accurate information, what do you do?

10 Do you have any questions that haven't been answered in this chapter? Where could you go to find the answer?

6

COMMUNITY AND RELATIONSHIP SAFETY

Shutterstock.com/Nicetoseeya

WORKSHEET 6.1 GO-TO PEOPLE

Page 229

The following table contains a mix of scenarios and go-to people. Consider the information provided and then fill in the blanks. An example has been given.

Scenario/situation	Go-to person	Reason for choice
You get a 'bad' test result in science.	Science teacher	
	Classmate	
You ... (add a scenario)	Parent	They offer support and would understand the situation and were going to be contacted by the transport officers anyway. I wanted to give them the heads-up and we have an open and honest relationship.
	Aunt	
You pass out at a mate's party and your mobile phone is stolen.	Your mate	
	School counsellor	

WORKSHEET 6.2 GOOD, BAD OR UGLY

Page 230

Getty Images/Ghislain & Marie David de Lossy

Consider the image above and think of three good outcomes, three bad outcomes and three ugly outcomes (these are worse than bad!) that might happen when these teens celebrate the end of Year 12 down at the local beach.

1 Good outcomes

2 Bad outcomes

3 Ugly outcomes

9780170465540

WORKSHEET 6.3 SAFE PARTYING

SB Page 234

A KWFL chart is a fun way of taking stock of 'where you are at' in certain areas.

1 Complete the following KWFL chart for safe partying.

Consideration	What you already Know	What you Want to know	How you will Find out	What you have Learnt
Laws associated with purchase or consumption of alcohol				
Parental responsibilities				
Police information and involvement				
Local council rules and support				
Protecting your house or place where the party is being held				
What to do if things go wrong				

2 Complete your research and present your information via a poster, tri-fold information brochure or web page.

Investigation skills

WORKSHEET 6.4 ALCOHOL-RELATED ISSUES IN AUSTRALIA

SB
Page 235

When does risk end, and recklessness begin? In Australia, 38% of people have been affected by alcohol-related violence. And just under a third (32%) of Australians believe that alcohol is the drug that causes the most harm. Risky, reckless behaviour can result in harm to yourself or others.

The table below shows the alcohol-related issues Australians are most concerned about over an eight-year period.

	2012 (%)	2013 (%)	2014 (%)	2015 (%)	2016 (%)	2017 (%)	2018 (%)	2019 (%)
Road traffic accidents	82	80	79	77	76	78	78	74
Violence	76	78	81	78	79	76	75	74
Child abuse and neglect	68	70	66	64	64	71	70	65
Crime	52	57	59	54	58	57	56	57
Health problems	62	62	52	51	53	53	57	54
Harm to unborn babies from exposure to alcohol in-utero	57	59	52	48	50	54	52	54
Lost productivity	27	31	21	22	24	29	27	25
Excessive noise around pubs and clubs	24	26	19	20	2	21	18	23
None of the above	2	4	3	5	5	3	5	5

Annual Alcohol Poll 2019: attitudes & behaviours, The Foundation for Alcohol Research and Education (FARE)

compare
to observe or note how things are similar or different

1 a **Compare** the alcohol-related issues concerning Australians in 2012 and 2019 by creating a graph in the space below.

9780170465540

b Do you believe there has been any significant change in the past 8 years? Use data to support your answer.

c **Propose** why there has/has not been any significant change.

propose to put forward (e.g. a point of view, idea, argument, suggestion) for consideration or action

AC

d Select one of the concerns from the table and design a campaign to reduce that alcohol-related issue.

2 Risky behaviour, such as drinking over the recommended amount, can lead to recklessness when your behaviour begins to harm yourself and others. Some of those harms are detailed in the table above. Use the scenarios below to **discuss** the impact of that reckless behaviour.

discuss to talk or write about a topic, taking into account different issues or ideas

AC

a You become a paraplegic after getting into a car with an unlicensed driver and being involved in a serious car crash. How would your life be affected?

b You were at a party and tried a substance because all of your friends were using this substance and you felt pressured to do so as well. You became addicted. How will your daily life be different now?

c You are at a party and someone passes out because they have had too much to drink and you want to call for an ambulance. However, your mates say, 'Don't call for an ambulance, the cops will be here in no time and the party will be shut down'. Conduct some research to find out if this is a fact or a myth/ non-truth, and summarise your findings.

3 Select three other sources of information and summarise their recommendations about teen drinking.

a

b

c

 9780170465540

WORKSHEET 6.5 BRAIN DEVELOPMENT

There is a great deal of research and information that links impaired brain development to alcohol and drug consumption during teenage years.

Areas of the brain

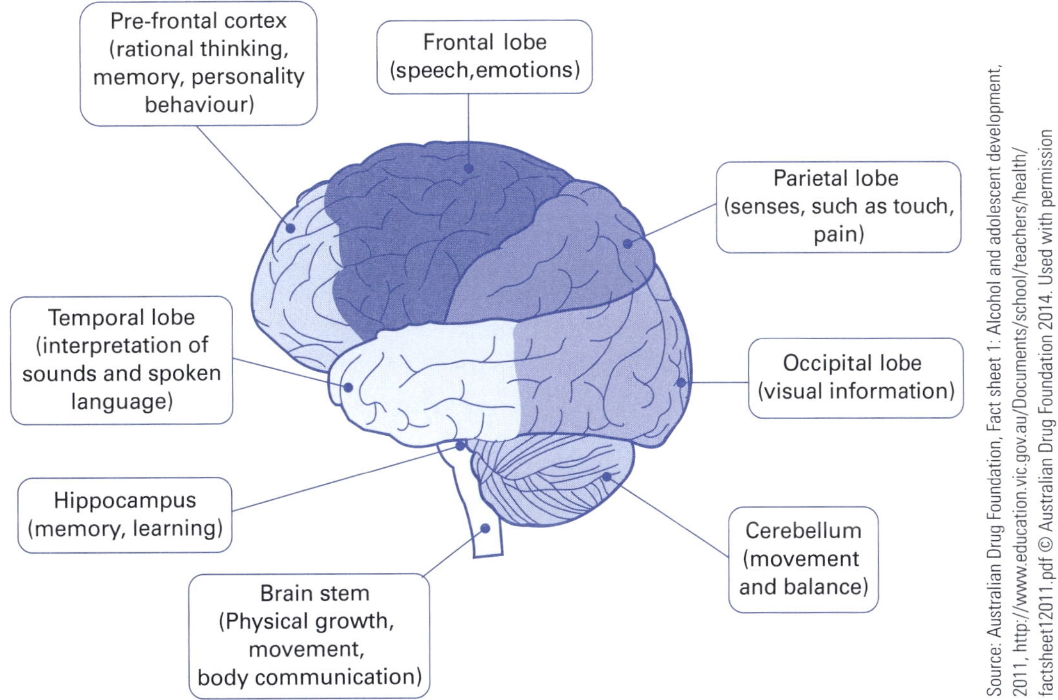

Source: Australian Drug Foundation, Fact sheet 1: Alcohol and adolescent development, 2011, http://www.education.vic.gov.au/Documents/school/teachers/health/factsheet12011.pdf © Australian Drug Foundation 2014. Used with permission

On the fishbone diagram below, list the likely future effects of the damage that alcohol can cause to each part of the brain indicated (e.g. one effect of damage to the hippocampus is impaired memory).

Likely future effects

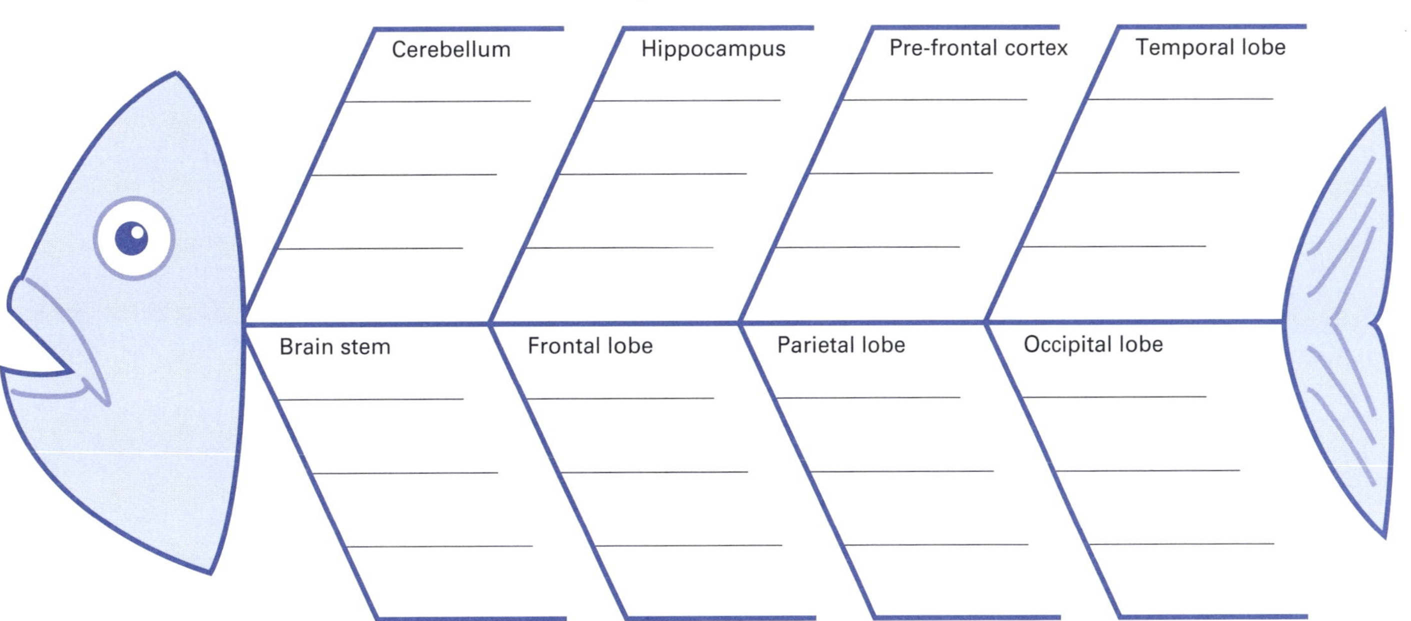

WORKSHEET 6.6 BEST CASE OR WORST CASE?

Page 239

Complete the following table, which considers different environments where your personal safety might be challenged. You may need to use your student book or research some of the safety information.

- In the first column, provide possible challenges.
- In the second column, provide two coping and helpful strategies (best-case scenarios).
- In the third column, provide two potential negative outcomes (worst-case scenarios).

Environment/setting	Situation	Best-case scenarios/ strategies	Worst-case scenarios/strategies
Using your mobile phone		1	1
		2	2
Open house party (parents away for the weekend)		1	1
		2	2
Walking to a venue		1	1
		2	2
Catching a train to the city to watch a band		1	1
		2	2
Accepting a drink from someone at a party		1	1
		2	2

WORKSHEET 6.7 ALTERNATIVE ENDINGS

1 With two or three other classmates, write a script that describes a street situation where potential conflict can arise and offer two alternative endings. One possibility results in a combination of verbal, physical and emotional abuse or violence. The other possibility results in the conflict being defused. The following template might help to organise your thoughts.

Page 240

Think about ...	Situation resulting in conflict	Additional information
What happened?		
What did you do?		
How did others around you act?		
What were you thinking?		
Did you think, then act?		
Were there winners and losers?		
How can everyone win out of the actions taken?		
	Situation defused	**Additional information**
What happened?		
What did you do?		
How did others around you act?		
What were you thinking?		
Did you think, then act?		
Were there winners and losers?		
How can everyone win out of the actions taken?		

2 The script should then be role-played for the rest of the class to view and comment on. In particular, what strategies were put in place that resulted in a positive ending or pathway?

Investigation skills

WORKSHEET 6.8 IMAGINE

SB Page 240

Conflict sometimes ends in violence, especially when alcohol is involved. Organisations such as Step Back Think or Coward's Punch Campaign have worked hard to raise awareness of the dangers associated with conflict.

The below data shows a summary of experiences of alcohol-related violence in Australia over a nine-year period.

	2012 (%)	2013 (%)	2014 (%)	2015 (%)	2016 (%)	2017 (%)	2018 (%)	2019 (%)	2020 (%)
Have been a victim of alcohol-related violence	14	18	19	14	16	19	21	18	18
Have had a family member or friend be a victim of alcohol-related violence	22	21	26	22	20	24	24	26	21
Neither of the above	69	68	63	70	71	65	63	62	60

Annual Alcohol Poll 2020: attitudes & behaviours, The Foundation for Alcohol Research and Education (FARE)

1 a Use the space below to represent the above data in a line chart. Hint: use different colours to plot the experiences.

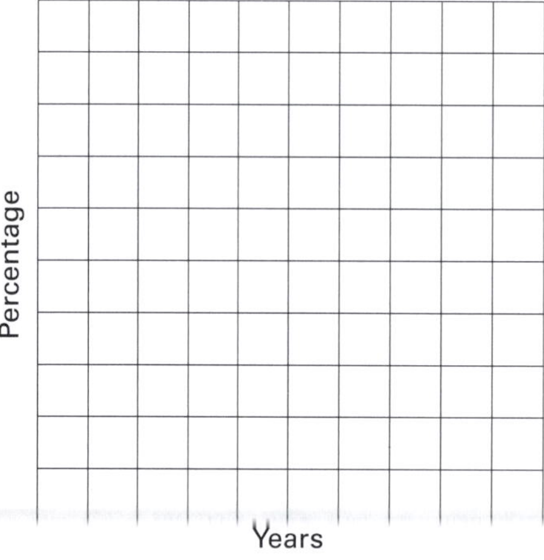

b **Describe** the trends over time you see with regards to alcohol-related violence.

____

____

____

____

____

describe to give an account of characteristics or features

The below table shows the ways under-18s have been harmed due to someone's drinking.

	2016 (%)	2017 (%)	2018 (%)	2019 (%)	2020 (%)
They have been verbally abused	13	12	13	11	14
They have been neglected in some way	11	9	9	8	11
They have been in a car with a driver who was over the legal blood alcohol limit	10	7	8	8	12
They have been physically abused	7	7	5	8	10
Child has been harmed or put at risk in some way	23	21	23	23	28
Child has not been harmed or put at risk in some way	77	79	77	77	72

Annual Alcohol Poll 2020: attitudes & behaviours, The Foundation for Alcohol Research and Education (FARE)

2 a Create a bar chart to **compare** 2016 data with 2020 data.

compare to observe or note how things are similar or different

AC

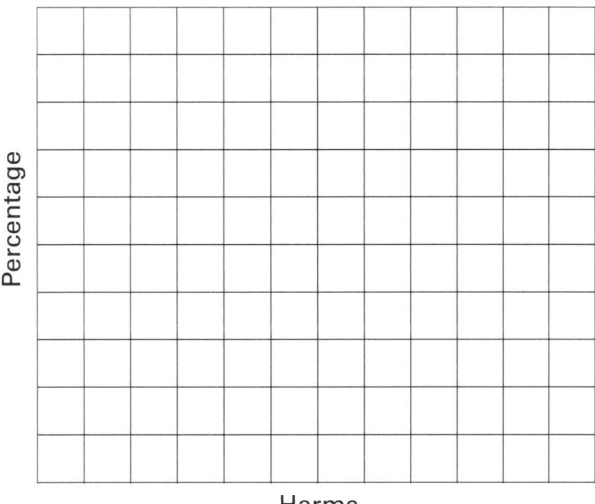

b Analyse this data by describing two trends you observe.

____

____

____

____

____

3 Imagine a man who was 'coward punched' while trying to shield his partner from harm and died in hospital the following day. **Discuss** how the following people would feel.

discuss to talk or write about a topic, taking into account different issues or ideas

a His partner

____

____

b His mates

c His four-year-old child

d His attacker

e His attacker's partner

f His attacker's mates

4 The data shows an increase in alcohol-related harm over the past 3–4 years, despite the efforts from campaigns and organisations such as Step Back Think and Coward's Punch Campaign. Suggest three strategies that campaigns like these could use to have more of an impact on reducing alcohol-related violence.

WORKSHEET 6.9 DRIVING ADVERTISEMENT

SB Pages 242–5

Young drivers make up a disproportionate number of people involved in car accidents. Many of these accidents happen because of inexperience, lack of concentration or distractions and risk-taking behaviours common among 18- to 25-year-old drivers.

1 In the space below, create a storyboard for a TV 'infomercial' raising awareness of the dangers associated with being an 18- to 25-year-old driver. In each of the 'scenes' or boxes, draw or sketch what you want to convey to the audience. In the row underneath, write any lines you want to be spoken or narrated in the infomercial.

Infomercial title:

Scene 1	Scene 2	Scene 3	Scene 4
Words/information	Words/information	Words/information	Words/information

Scene 5	Scene 6	Scene 7	Scene 8
Words/information	Words/information	Words/information	Words/information

WORKSHEET 6.9 CONTINUED

2 Advertisers often use 'shock tactics' to highlight the dangers associated with certain behaviours in an effort to get people to stop and think about their behaviour and to consider how they could improve the situation. Examples of shock tactics include the images on cigarette packets, TV commercials showing outcomes of people who aren't 'sun-smart' and, more recently, the effects of poor diet and lack of exercise.

discuss
to talk or write about a topic, taking into account different issues or ideas

a Generalise if you believe these shock tactics are changing people's behaviours and improving their health? **Discuss**.

b Imagine you have been asked to design a shock tactic campaign to change the behaviour of young drivers. Briefly outline three different shock tactics you would focus on in your campaign.

Shock tactic 1

Shock tactic 2

Shock tactic 3

WORKSHEET 6.10 HOONING

6

SB
Pages 252–6

Use the internet to find articles about 'hooning', the negative effects of such behaviour and the consequences of being caught.

1 Contrast the consequences of hooning in your state compared with at least two other states or territories.

2 Identify and justify which state or territory's anti-hoon laws you think are the most effective in eliminating this type of behaviour on our roads.

3 Summarise the findings of at least two articles about hooning and present this as a 200-word report.

WORKSHEET 6.11 ANTI-DISCRIMINATION LAWS

Page 250

Use the table template below for both activities.

1 Search online to find anti-discrimination laws that exist in your state. Summarise these in point form.

2 Investigate the policies your school has to protect students and staff from discrimination. Summarise these in point form.

Types of discrimination	State/national laws	School policies

WORKSHEET 6.12 IT'S YOUR RIGHT TO SAY NO

Page 253

Sometimes you need to say 'no' and assert your right not to be part of an activity. Remember, it is your right to say no.

Many myths exist around saying no. Some of these are:

- saying 'no' is rude or aggressive
- saying 'no' is unkind, uncaring and inconsiderate
- saying 'no' is hurtful and will upset others
- saying 'no' will mean they won't like me anymore, and will never ask me again.

1 Recall other reasons you have experienced those around you avoiding to say 'no'.

____

____

____

____

____

____

2 For each of the following scenarios, provide an assertive statement that distinguishes between fact and opinion. For example, 'You might think that ... but actually I feel ...' You could also put yourself at the centre of what you say by using 'I' statements, such as 'I think that ...'.

SCENARIO 1

You are in a new relationship and have been 'going out' for a couple of weeks. The other person wants to get a little more intimate but you don't want this. They say, 'if you really loved me, you would'.

____

____

____

____

____

____

____

____

____

SCENARIO 2

You and your mates are part of a school sports team. Someone outside your friendship group 'stuffs up' at an important stage of the match, costing your team the game. After school your mates start to tease the person, calling them a 'loser'. This makes you feel very uncomfortable.

SCENARIO 3

You have invited some friends to your house before going to a party. Your parents aren't home, and one of your friends grabs a bottle of vodka from a cupboard. They have a swig before passing the bottle to you. Everyone starts chanting 'scull' and 'chicken', making you feel very stressed and uncomfortable because you don't want to have a drink.

9780170465540

SCENARIO 4

You are already feeling bad about skipping school and spending the day with friends at the local shopping mall. The day has been pretty good until you are in a shop and your friend takes a t-shirt from a rack and stuffs it under their jumper. You know this is stealing and against the law, but you feel you are being pressured to do the same by the look on your friend's face.

3 **Discuss** a situation where you have stepped in for a friend and given them support, which they otherwise would not have received. What might have happened if you weren't there, or had not had the courage to be assertive on your friend's behalf?

discuss
to talk or write about a topic, taking into account different issues or ideas

AC

WORKSHEET 6.13 FIRST AID

Page 257

Go to the ABC Health & Wellbeing website and search for the First Aid Quiz, which has 10 multiple-choice questions.

1 Answer the multiple-choice questions related to 10 emergency scenarios. You can check your answers immediately.

2 Create four common emergency scenarios, two based at school and two outside school. Once you have completed them, show your scenarios to a classmate and ask them to try to decide the first aid steps they would put in place.

FIRST AID

Scenario 1

Scenario 2

Scenario 3

Scenario 4

9780170465540

WORKSHEET 6.14 MAKE A BROCHURE

Pages 258–9

Concussion can occur from a knock to the head while playing sport, falling awkwardly off a bike and hitting your head or being struck on the head by an object. People do not necessarily need to be 'knocked out' to suffer from concussion.

1 Design a tri-fold pamphlet that focuses on concussion – its causes, signs and symptoms, and suggested treatment.

   a Use a blank piece of paper, folded using the instructions below to brainstorm and plan your ideas. The first three columns represent the front of the pamphlet and the second three columns represent the back.

   b Once you have finished your plan/sketch/information, swap with a classmate and give each other some constructive criticism and suggestions for improvements. If you agree with the suggestions, make the changes. If you do not believe the suggestions will improve your pamphlet, respectfully argue your point of view.

   c Using a blank sheet of paper, create your tri-fold pamphlet. Remember to acknowledge your sources of information (if relevant).

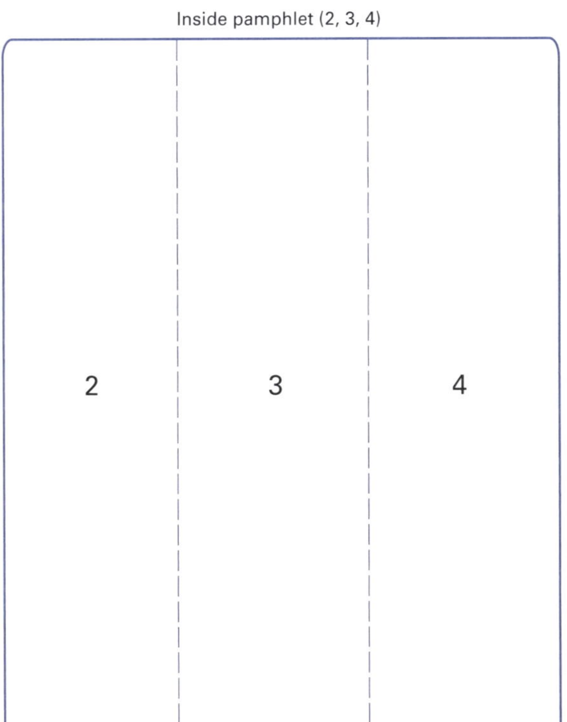

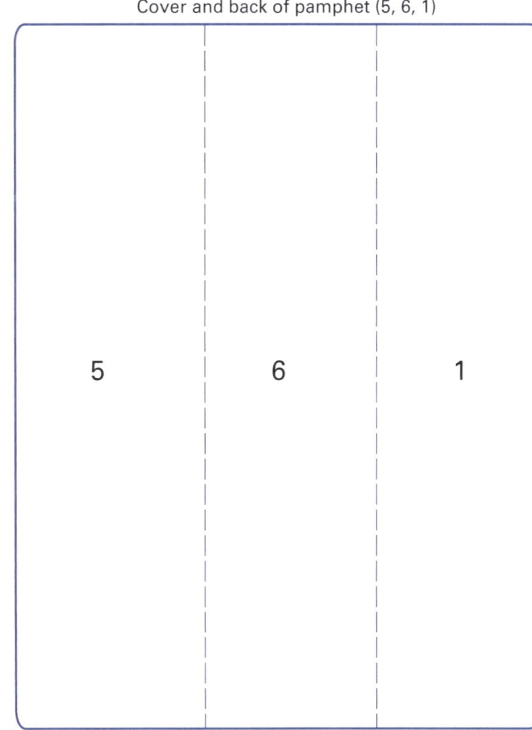

CHAPTER 6 REVIEW

Self-reflection at the end of your work is a powerful way to gauge what you now know, what you enjoyed and what you need to find out.

Complete the following statements relating to the work covered in this chapter.

1 I now know that ...

____

____

____

____

____

2 I really didn't understand ...

____

____

____

____

____

3 The most interesting things I learnt were ...

____

____

____

____

____

4 I enjoyed working with ...

____

____

____

____

____

9780170465540

5 The next group that this is presented to should ...

6 I would like to find out more about ...

7 My favourite activity was ...

8 I was engaged when ...

9 It may have been good to also consider ...

7

BEING ACTIVE AND ADVENTUROUS OUTDOORS

Shutterstock.com/Nicetoseeya

WORKSHEET 7.1 NOW AND THEN

Using the Venn diagram below, compare and contrast a sport (such as AFL) today with the same sport during the 1800s. Remember to include what is different and what is similar about the sport.

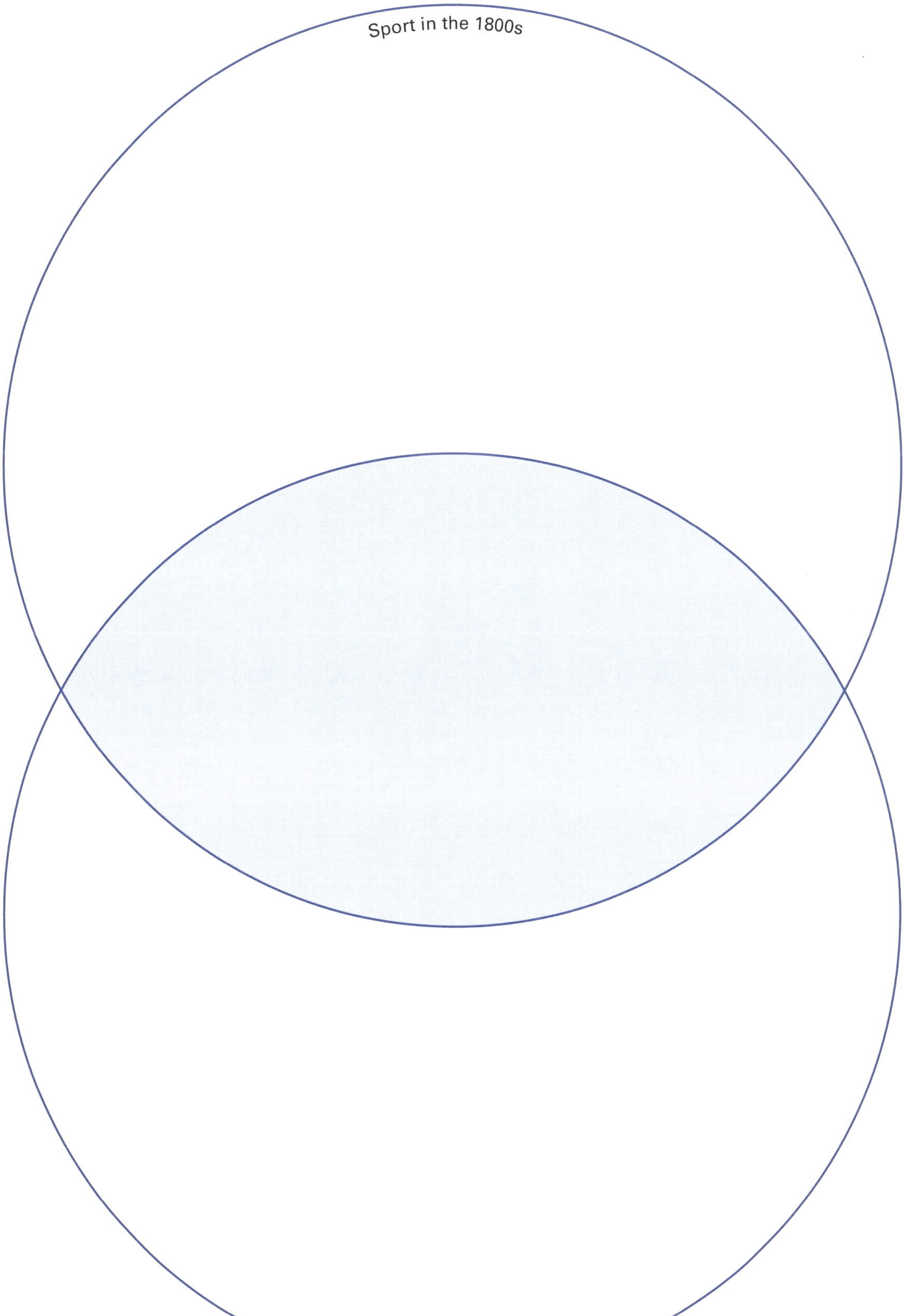

Sport in the 1800s

Sport today

Page 274

WORKSHEET 7.2 MY OPINION

justify
to show how an argument or conclusion is right or reasonable

How strongly do you agree with the statements below? For each statement, place its number where you feel it should fit on the continuum. There are no right or wrong answers. Use the lines provided to write a **justification** of your answer.

Strongly disagree	Disagree	Neither agree nor disagree	Agree	Strongly agree

1 Every student at school should experience some form of outdoor recreation on a regular basis.

____

____

____

2 The benefits of outdoor recreation are not as important as participating in sport.

____

____

____

3 To be involved in outdoor recreation is expensive and requires a high level of skill.

____

____

____

4 Outdoor recreation is only important for people who want to be challenged physically.

____

____

____

5 If more people were involved in outdoor recreation, there would be more respect for nature and the environment.

____

____

____

WORKSHEET 7.3 THE BENEFITS OF OUTDOOR EDUCATION YOUTH PROGRAMS

SB Page 275

The Outdoor Youth Programs Research Alliance (OYPRA) conducted research in Australia to explore the benefits to young people of participating in outdoor, camping and nature-based programs. They surveyed around 250 outdoor education providers across Australia about their programs, the activities they offered, and the participants' involvement. The below graphs and statistics detail some of its results. Analyse the figures below and then answer the following questions.

Figure 1 This figure shows how much time (in days) was spent on each adventure activity across the outdoor education programs

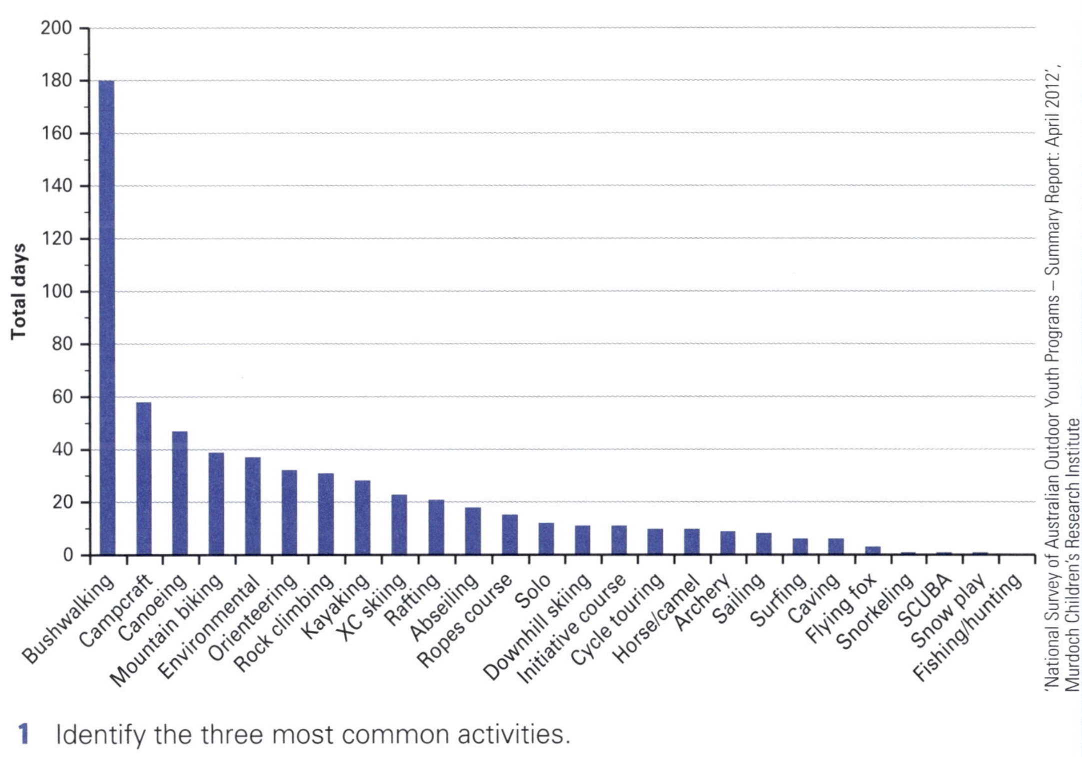

'National Survey of Australian Outdoor Youth Programs – Summary Report: April 2012', Murdoch Children's Research Institute

1 Identify the three most common activities.

____

____

____

2 Propose why you believe these are the most common.

____

____

____

____

3 Survey your class members to see who has participated in the activities mentioned. You can use the table below to tally your results.

Activity	Tally	Total
Bushwalking		
Campcraft		
Canoeing		
Mountain biking		
Environmental		
Orienteering		
Rock climbing		
Kayaking		
XC skiing		
Rafting		
Abseiling		
Ropes course		
Solo		
Downhill skiing		
Initiative course		
Cycle touring		
Horse/camel		
Archery		
Sailing		
Surfing		
Caving		
Flying fox		
Snorkelling		
SCUBA		
Snow play		
Fishing/hunting		

4 Summarise the top three activities your class has participated in.

___

___

___

9780170465540

Figure 2 This figure shows the intended goals of the programs and the number of young people the programs believed experienced that goal.

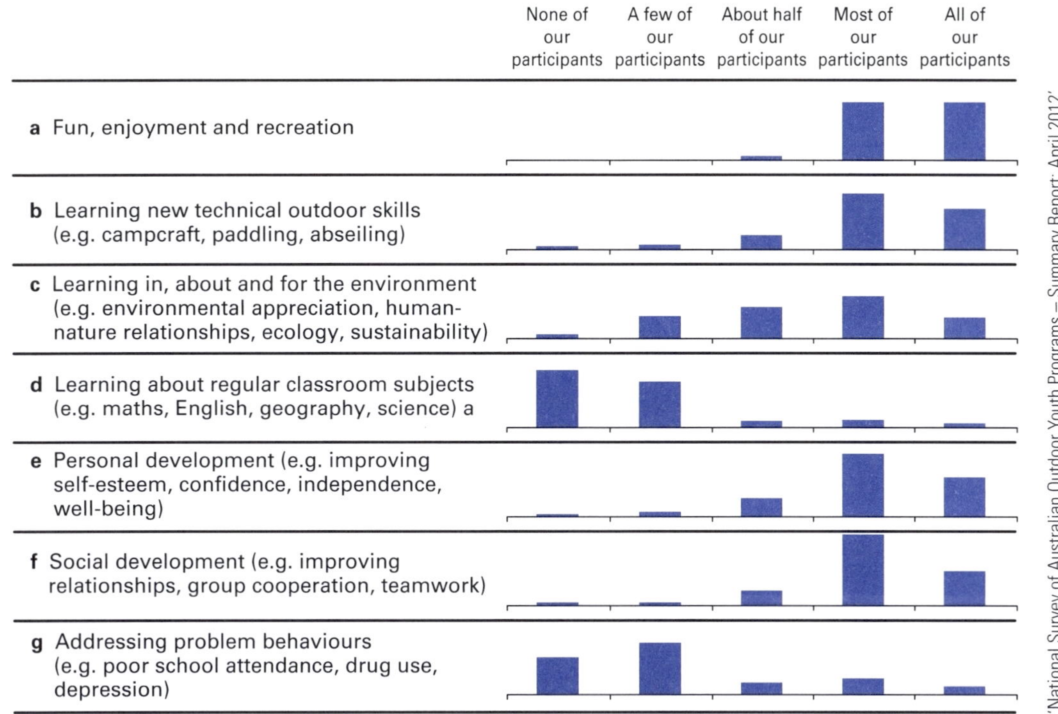

'National Survey of Australian Outdoor Youth Programs – Summary Report: April 2012', Murdoch Children's Research Institute

Figure 3 This figure indicates the research methods used by the staff survey responders to determine whether the above goals had been met.

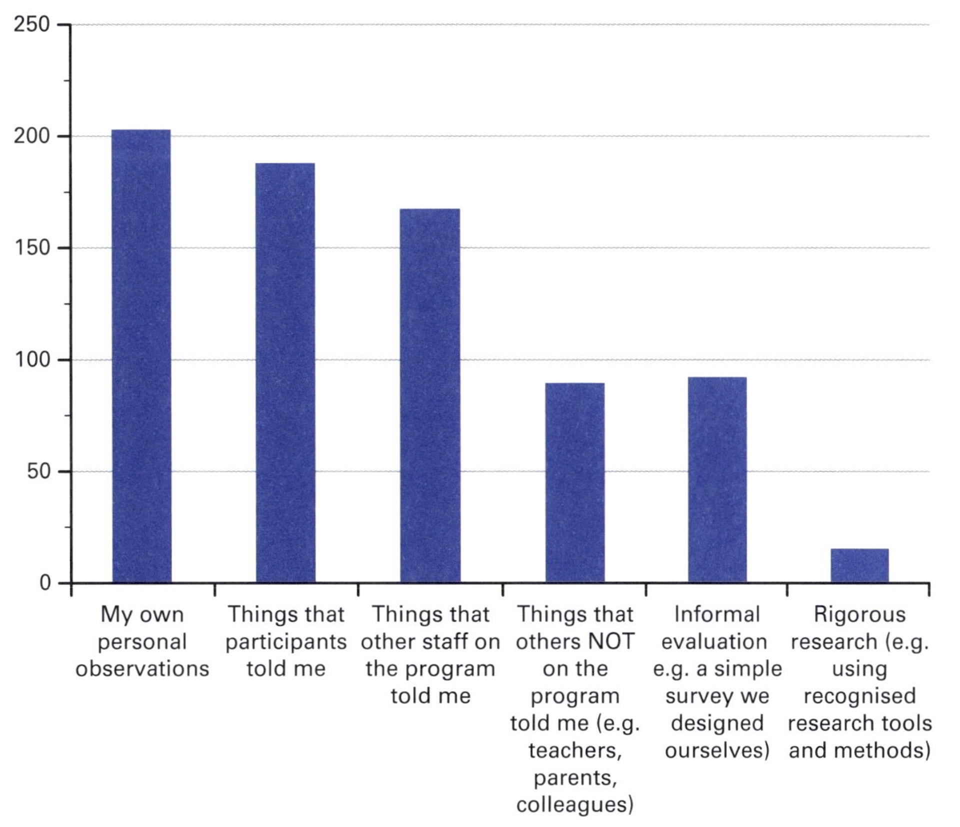

'National Survey of Australian Outdoor Youth Programs – Summary Report: April 2012', Murdoch Children's Research Institute

5 Analyse the research methods in Figure 3 above. Do you think the results in Figure 2 regarding meeting the goals of the youth programs can be confidently assured given the research methods chosen in Figure 3? Why/why not?

6 **Discuss** which of these you think is the best method for achieving the OYPRA's research goal. Why?

discuss
to talk or write about a topic, including a range of arguments, factors or hypotheses; consider, taking into account different issues and ideas, points for and/or against, and supporting opinions or conclusions with evidence

9780170465540

WORKSHEET 7.3 CONTINUED

7 Survey your class to see if the goals listed above were met in the last outdoor recreation activity you completed. You can use the below table to help tally your results.

Goal	Yes – goal was met		No – goal was not met	
	Tally	Total	Tally	Total
Fun, enjoyment, recreation				
Learning new technical outdoor skills				
Learning in, about and for the environment				
Learning about regular classroom subjects				
Personal development (e.g. improving self-esteem, well-being, confidence)				
Social development (e.g. improving cooperation, teamwork)				
Addressing problem behaviours (e.g. poor attendance, drug abuse)				

8 From your own research, summarise the main findings about participation in outdoor recreation.

9780170465540

WORKSHEET 7.4 MEDIA ANALYSIS

Read the article below, then answer the six focus questions to summarise your understanding of how the media can influence readers about outdoor recreation safety.

CASE STUDY

Sydney canyoner in Blue Mountains rescue

A Sydney canyoner was flown out of the Blue Mountains today with a fractured ankle amid police pleas for bushwalkers to stop taking unnecessary risks.

Police said the Bellevue Hill woman was canyoning with a group of friends in the Whungee Wheengee Canyon, near Mount Wilson, close to where 15-year-old Nick Delaney died in a suspected rock collapse last month. The 25-year-old fractured her ankle, Detective Inspector Michael Bostock, said: '[The woman's] party had recorded their trip with police and they took a beacon with them, so with that we were able to know who they were and exactly where they were'.

A member of the woman's group stayed with her while the others trekked out and alerted the authorities. Emergency services tried to retrieve the group yesterday but darkness and bad weather forced them to wait until today. The woman was finally flown out of the canyon about 6.30 a.m. and taken to Westmead Hospital, where she remained in a stable condition.

The incident comes a day after police issued three canyoners with court attendance notices after they allegedly ignored locked gates and warning signs against going canyoning in the Blue Mountains in heavy rain. The group became caught on a rock ledge after water levels in the Grand Canyon rose rapidly under a torrential downpour. A police rescue unit, plus paramedics and National Parks workers had to rescue the trio.

National Parks regional manager Geoff Luscombe said the trio's actions were 'reckless' and endangered their own lives as well as those of their rescuers. They will appear at Katoomba Local Court in April for the offence of engaging in an activity in a national park that causes risk to safety.

Blue Mountains police said they were sick of rescuing bushwalkers and canyoners who did not take the proper precautions.

As of Sunday, police rescue officers had conducted 53 operations across the area since January 1. About 80 per cent of those were for people reported missing in the bush. 'It is frustrating to the extreme that people demonstrate a total disregard for their own personal safety, and as a result the potential danger to police officers and other emergency services personnel who have to come to their rescue,' Detective Inspector Bostock said.

WORKSHEET 7.4 CONTINUED

1 a **Who** is this article aimed at?

b **What** positive and negative aspects are highlighted in the article?

2 **Why** do you think people demonstrate such disregard for their own personal safety and the safety of others?

3 **Where** else can safety issues be raised, other than in the media, to highlight a serious problem?

4 **How** might the National Parks and rescue personnel provide solutions to this ongoing issue?

9780170465540

5 In the space below, put together a storyboard for a TV infomercial raising awareness of the dangers associated with participating in various activities in the outdoors. In each of the 'scenes' or boxes, draw or sketch what you want to convey to the audience. In the row underneath, write any lines you want to be spoken or narrated in the infomercial.

Infomercial title:			
Scene 1	Scene 2	Scene 3	Scene 4
Words/information	Words/information	Words/information	Words/information
Scene 5	Scene 6	Scene 7	Scene 8
Words/information	Words/information	Words/information	Words/information

WORKSHEET 7.4 CONTINUED

6 Advertisers often use 'shock tactics' to highlight the dangers associated with certain behaviours in an effort to get people to stop and think about what might happen when they take unnecessary risks. There are also media/community safety campaigns; for example, driving in flood waters. The Queensland Fire and Rescue Service website has a Community Safety page. Search for 'If it's flooded, forget it', and watch the short video clip on flood and swift water safety.

discuss to talk or write about a topic, taking into account different issues or ideas

a In your opinion, **discuss** if you believe these shock tactics or other media campaigns are changing people's behaviours.

____

____

____

____

____

b Imagine you have been asked to design a shock campaign to change people's complacency in relation to risky behaviour outdoors. Briefly outline three different shock tactics you would include in your campaign.

Shock tactic 1

____

____

____

____

Shock tactic 2

____

____

____

____

Shock tactic 3

____

____

____

____

WORKSHEET 7.5 CREATE A COMIC STRIP

Pages 280–1

Developing as a team or group can take time. Groups go through a series of stages, including formation, difficulties and working together.

Alamy Stock Photo/PhotoAlto

Formation

iStock.com/DMEPhotography

Difficulties

Newspix/Justin Lloyd

Working together

1 Create a comic strip, based on these stages, about a group of students that come together to complete a physical challenge in the outdoors. You are to create the situations, the characters, the activities and the speech bubbles.

a Use the storyboard below to create a rough outline of your comic strip.

b Create your comic strip, using either a free online comic strip maker, or draw it on butcher's paper.

Scene 1	Scene 2	Scene 3	Scene 4
Words/information	**Words/information**	**Words/information**	**Words/information**
Scene 5	**Scene 6**	**Scene 7**	**Scene 8**
Words/information	**Words/information**	**Words/information**	**Words/information**

WORKSHEET 7.6 ROLE-PLAY

Page 281

There are three recognised styles of leadership that people can use, depending on the circumstances. Often people use a combination of these styles.

Thinkstock/Comstock Images

An autocratic leader makes all the decisions.

Dreamstime/Konstantin Chagin

A democratic leader allows team members to have input to the decisions.

Getty Images/Kevin Dodge

Laissez faire leadership allows all the team members to make decisions.

1 Create a scenario in the outdoors that requires at least two of these styles of leadership to deal effectively with an issue that could otherwise develop into a serious problem.

You are then going to role-play the scenario to another pair and they will need to explain and justify why that type of leadership is required in that particular situation. Write out the scenario first, and if it is too difficult to act out, then the written scenario can be used.

Alternatively, you can reverse this activity so that the style of leadership is role-played and partners have to justify which leadership style was represented and why.

A rough outline of the scenario can be written in the table below with a justification for why that leadership style is best suited to the situation.

Aspect of the scenario	Leadership style	Justification of leadership style

WORKSHEET 7.7 'SPEED DATING'

SB
Page 284

This activity involves researching an activity that takes place in the outdoors; it could be an established activity or a relatively new activity (such as geocaching or dune surfing).

1 Gather as much information as you can about the activity and then summarise the key points into a one- to two-minute presentation. The presentation can incorporate multimedia (with video/photos) or it can be an online format. It must involve you speaking, so that you are required to learn and present the information you have researched. The outline below might be useful in planning the information in your presentation.

2 Present your activity in the form of a 'speed dating' session where half the class move from one presentation to another after a set period of time.

Name of activity:	
What year and in what country was it established?	
What are the main skills involved?	
What equipment is required?	
What aspects of safety must be addressed?	
What age group participates in this activity?	
How popular is the activity?	
Where can you participate in this activity? Is there a website?	
What risks are involved?	
Does the activity have potential to damage the environment? If so, how might this be minimised?	
Other interesting facts that you have found out about the activity.	

WORKSHEET 7.8 BLINDFOLD TENT GAME

Effective communication is essential for many aspects of living; it becomes even more important if you are in a leadership position or participating in an outdoor/adventurous activity.

For this task you will need a tent – still packed, and a blindfold. Ideally the class will have access to more than one tent, in which case there can be multiple groups working.

In this activity you will work in groups of three – one person blindfolded, one communicating and one observing. The challenge is for the blindfolded person to erect the tent by listening to the communicator giving as precise instructions as possible. The observer evaluates the person's ability to communicate effectively to the blindfolded person.

You should allow 10 minutes and then provide feedback to the communicator. The following checklist could be used to offer feedback.

Aspects of communication	Tick		How to improve – if required
	Yes	No	
Spoken word was clear and audible.			
Instructions offered were simple to follow.			
Alternative suggestions were used if instructions were not initially successful.			
Patience was evident.			

WORKSHEET 7.9 RECOMMENDED EQUIPMENT FOR BUSHWALKING

Page 304

The equipment required and the appropriate 'type' of equipment used is dependent on the specific *context* of the activity.

Equipment used for bushwalking may include but is not limited to:

Emergency/rescue

- Documentation
- Emergency communication equipment
- First-aid kit in waterproof storage
- A waterproof method of storing and carrying documentation and communications equipment
- Specific activity context equipment required
- Emergency shelter where appropriate for the context
- Emergency equipment to keep a patient warm (e.g. mat, sleeping bag) where appropriate for the context
- Signalling device(s) (e.g. mirror, flares)

Activity leaders

- Communications equipment (standard communication rather than emergency communication where this differs) and spare batteries or backup 'power banks'
- Relevant maps and navigation information
- A waterproof method of storing and carrying maps and navigation information
- Compass and/or other navigation aids (e.g. GPS)
- Pen/pencil and blank writing paper
- Watch or equipment suitable to tell and measure time for first-aid purposes
- Head torch and spare batteries
- Appropriate spare equipment
- Appropriate repair equipment
- Same as for participant

Participant

- Personal medications (including for asthma and anaphylaxis)
- Personal hygiene requirements
- Whistle
- Strong backpack, suitably sized and adjusted
- Waterproof pack liner
- Water containers
- Jumpers
- Thermals
- Beanie or balaclava

- Sun hat
- Raincoat suitable for the environment
- Overpants
- Footwear suitable for the conditions
- Suitable socks
- Shirt with collar and preferably long sleeves
- Strong shorts or trousers
- Underwear
- Gloves
- Light garden gloves
- Handkerchief
- Pocket knife
- Sock covers or gaiters
- Sunglasses
- Spare prescription glasses
- Walking pole(s)
- Sit mat
- High-visibility vest
- Sunscreen

Group

- Trowel for toileting
- Toilet paper
- Hand sanitiser
- Water purification system
- Repair kit
- Food for duration plus spare
- Rubbish bags
- Multi-tool with knife
- Sunscreen
- Insect repellent

Repair equipment

- Buckles
- Needle and thread
- Duct tape
- Multi tool
- Multi-purpose glue
- Tent pole repair tubes
- Sleeping mat repair kit

9780170465540

Recommendations for a first-aid kit contents may include but are not limited to:

Group first-aid kit

- Waterproof storage
- Disposable gloves
- Resuscitation mask or shield
- Conforming bandages
- Elastic bandages
- Triangular bandages
- Non-adhesive sterile dressings
- Waterproof sterile dressings
- Saline solution
- Antiseptic solution
- Safety pins
- Adhesive strapping tape
- Adhesive medical dressings
- Blister dressings
- Hand sanitiser
- Scissors
- Notebook and pencil
- Portable splint
- Support bandages for knees and elbows
- Space blanket
- Asthma medication
- Asthma spacer
- Anaphylaxis medication auto-injector
- Antihistamine medication
- Cream for chaffing
- Pain relief
- Rehydration solution
- Tweezers
- Splinter probes
- Booklet/notes on first aid treatment where necessary

Participant first-aid kit

- Personal medications
- Strapping tape
- Blister dressings
- Adhesive medical dressing (small)
- Sunscreen
- Insect repellent

Adapted from 'Bushwalking – Australian Adventure Activity Good Practice Guide', Outdoor Council of Australia, 2019, and 'Core – Australian Adventure Activity Good Practice Guide', Outdoor Council of Australia, 2019

CHAPTER 7 REVIEW

Reflect upon your learning in this chapter by completing the following sentences.

1 One aspect of the chapter I found interesting was ...

2 I thought ...

3 I now know that ...

4 I was surprised that ...

5 I enjoyed ...

6 I will always remember ...

7 Two skills I have learnt are ...

8 I still wonder ...

9 I think we learnt about this topic because ...

8

PARTICIPATING AND PERFORMING IN GAMES AND SPORTS

Shutterstock.com/Nicetoseeya

WORKSHEET 8.1 GAMES

CULTURAL GAMES

describe
to give an account of characteristics or features

AC

1 Games and sports are played by First Nations Peoples around the world. For example, Australian First Nations Peoples have a range of games including marngrook (Indigenous Australian football game), meetcha boma (a hockey game) and other cultural games. Lacrosse is a traditional game from Canada and North America, and a game from South-East Asia is sepak takraw (or kick volleyball).

Research a cultural game from a country of your choice and **describe** the game in enough detail so that someone would be able to play the game from reading your description. Use the space below to include some drawings to help people understand how to play.

9780170465540

RULES

Games are recreational activities in which you solve problems while obeying a set of rules. Sports are games that are competitive, use movement skills, have a wide following, and usually have an organising body.

In games and sports, rules present challenges to be overcome.

- Primary rules determine how a game is played and won. If the primary rules are changed, the game becomes a different game.
- Secondary rules are those that can be modified without changing the nature of the game.

2 Identify whether the following rule changes would be changes to the primary or secondary rules in hockey. The first one has been done for you.

	Primary rule	Secondary rule
Using a smaller goal		✓
Playing five a side		
Playing on a smaller pitch		
Allowing players to kick the ball on purpose		
Allowing players to pick the ball up with their hands		

discuss to talk or write about a topic, taking into account different issues or ideas

3 **Discuss** with a partner why you decided each rule change was a change in the primary or secondary rules. You can make some notes in the space below.

____

____

____

____

____

____

____

____

WORKSHEET 8.2 TYPES OF GAMES AND SPORTS

Pages
317–19

Games and sports can be classified according to the interactions between the players or according to their tactical similarities.

INTERACTION BETWEEN PLAYERS

Some games and sports rely much more on cooperation and teamwork than others.

1 Classify different types of games and sports by placing the following games and sports on the continuum, from no interaction to high levels of interaction. Three sports have been entered for you.

- Netball
- Gymnastics
- Football
- Hockey
- Golf
- Martial arts
- Cricket
- Tennis
- Lacrosse
- Triathlon
- Skateboarding
- Figure skating
- Softball
- Lawn bowls

Individual sports
No interaction
between players

athletics

Moderate interaction
between players

baseball

Team sports
High interaction
between players

soccer

TACTICAL SIMILARITIES

There are four categories of games based on tactical similarities:

- **invasion (or territorial) games**, in which opposing teams attempt to invade their opponent's territory to score points
- **net/wall games**, in which players aim to send an object into an opponent's area so that the opponent cannot return it
- **striking/fielding games,** in which there is a batting team that tries to score as many runs as possible
- **target games,** in which the player's aim is to hit a target with an object, or come as close as possible to it.

2 Classify different types of games and sports as invasion, net/wall, striking/fielding, or target games (if you don't know what the game or sport is, look it up online before classifying it). The first one (croquet) has been completed for you.

Game	Classification	Game	Classification
Croquet	*Target*	Table tennis	
Baseball		Squash	
Soccer		Gridiron	
Volleyball		Ultimate disc	
Bocce		Tennis	
Sepak takraw		Tenpin bowling	
Rounders		Bat tennis	
Netball		Racquetball	
Football		Hockey	
Down ball		Badminton	
Cricket		Newcomb	
Curling		Softball	
Korfball		Kickball	
Archery		Lacrosse	
Rugby		Golf	
T-ball		Lawn bowls	

WORKSHEET 8.3 SKILL IN GAMES AND SPORTS

Pages 320–1

Skilled players can combine moving effectively with making decisions about what to do next. They have well-developed movement skills and tactical skills.

describe to give an account of characteristics or features

1 Identify your favourite player in a sport of your choice.

a Why are they such a good player and what makes them stand out for you?

b **Describe** three movement skills and three tactical skills they perform.

Sportsperson:

Sport:

Movement skills	Tactical skills
e.g. accurate passing	e.g. always seems to be in the right position to get the ball
1	1
2	2
3	3

2 Think about your own performance in that sport.

What three movement and three tactical skills do you think you need to work on to improve your performance?

Sport:

Movement skills	Tactical skills
1	1
2	2
3	3

WORKSHEET 8.4 MOVEMENT SKILLS

SB
Pages 320–4

Fundamental movement skills are the foundations that help you build more advanced and specialised movement skills to then apply to a specific sport.

1 For each of the following fundamental movement skills, identify a sport and a specialised movement skill in that sport that is based on the fundamental movement skill. The first one has been done for you.

Fundamental movement skill	Sport	Specialised movement skill
Throwing	*Tennis*	*Serve*
Catching		
Striking (hitting)		
Kicking		
Balancing		
Agility		
Running		
Jumping		

Movement skills can be classified as:

- **gross** (use large muscle groups for power) or **fine** (use smaller muscle groups for precision)
- **discrete** (a distinct movement with clear beginning and end), **serial** (a series of specific movements linked together) or **continuous** (a repetitive or rhythmical movement with no clear beginning or end)
- **open** (reacting and adapting to an unpredictable environment) or **closed** (consistent movement in a predictable environment).

WORKSHEET 8.4 CONTINUED

2 Match each of the listed skills according to the three classification types above. That is, for each skill on the left, you make three lines. The first has been done for you.

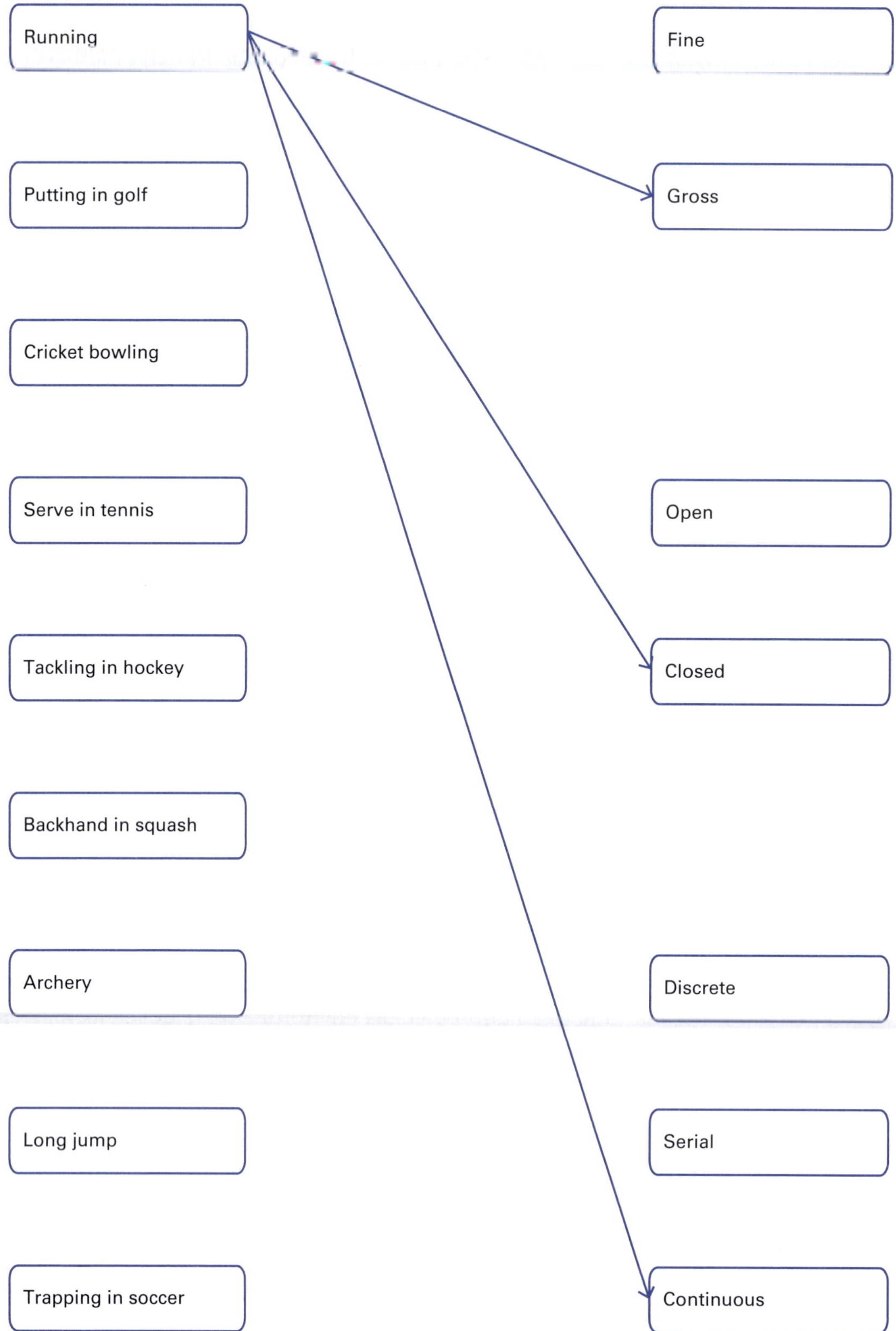

WORKSHEET 8.5 TACTICAL SKILLS

Pages 324–5

Tactical skills (tactics, strategies and game plans) are used in games to gain an advantage over the opponent.

1 What are the main tactics and strategies to improve performance in the games listed in the table?

Game	Main tactics and strategies
Netball (invasion game)	Creating space in attack; denying space in defence; defending the goal
Soccer (invasion game)	
Tennis (net/wall game)	
Softball (striking/fielding game)	
Lawn bowls (target game)	

2 This activity involves playing different games and comparing the tactical skills in those games.

a As a class, choose an invasion game to play. Play this game, then discuss the main tactical skills that you used. List them below.

____

____

____

____

____

____

WORKSHEET 8.5 CONTINUED

b Choose another invasion game to play. While you are playing, distinguish the main tactical skills that are being used and compare these with the tactical skills that you used in the other invasion game.

i What are the similarities between the two invasion games you played and what are the differences?

____

____

____

____

____

ii Why might these similarities and differences exist?

____

____

____

____

____

c Choose a net/wall game. Play this game, and as you are playing, distinguish the main tactical skills that are being used and compare these with the tactical skills you used in the invasion games.

i What are the similarities between the net/wall game and the invasion games you played? What are the differences?

____

____

____

____

____

ii Why might these similarities and differences exist?

____

____

____

____

____

9780170465540

Investigation skills

WORKSHEET 8.6 ANALYSING YOUR SKILLS

Page 329

Performance analysis helps to improve performance in games and sports. There are two types of performance analysis:

- **Motion analysis** involves observing the movement skill to analyse technique.
- **Games analysis** is used to assess the use and performance of movement skills and tactical skills within the game.

1 Analyse your performance using motion analysis. Choose a movement skill from a game or sport and analyse your performance and that of a partner. When performing the analysis make sure that you follow all the steps necessary for motion analysis. You might video the movement skill performance and use video analysis software or an analysis app.

a Use the table here to help you in your analysis:

Movement skill:		
Steps		**Notes on the movement skill**
Step 1	**Preparation and planning** What are the key elements of the skill?	
Step 2	**Observation** How will you observe the skill being performed?	
Step 3	**Detection and evaluation** Is the skill being performed correctly? Are there errors in the performance?	
Step 4	**Diagnosis** What is the cause of the error?	
Step 5	**Correcting errors** How can you help the performer to correct the error?	

b Once you have analysed the movement skill, discuss how to improve your performance and provide feedback to your partner about their movement skill.

c Develop a practice activity to improve your movement skill and practise the skill.

d After practising the skill, perform the skill analysis again and see if your practice has been effective at improving your performance and that of your partner.

WORKSHEET 8.6 CONTINUED

Investigation skills

2 Pick a sport and list five skills that a performance analyst could code to develop a games analysis in that sport.

Sport	
Skill 1	
Skill 2	
Skill 3	
Skill 4	
Skill 5	

The following table presents total kicking data from six elite Australian Football League (AFL) games. The data is for both teams combined and is separated into effective and ineffective kicks. It is coded for effectiveness, distance, kick type, region of ground, physical pressure and kicking foot.

Kick coding		Effectiveness			
		Effective kicks		Ineffective kicks	
		Number	%	Number	%
Kick distance	Short (less than 20 m)	629	79.6	161	20.4
	Medium (20–40 m)	413	58.9	288	41.1
	Long (more than 40 m)	426	69.7	185	30.3
Kick type	Set kick (from free kick or mark)	744	82.5	158	17.5
	General kick (in play)	631	63.5	363	36.5
	Kick from a contest (opposition close)	93	45.1	113	54.9
Region of the ground	Forward	337	60.7	218	39.3
	Mid-forward	378	58.6	267	41.4
	Mid-defence	442	77.4	129	22.6
	Defence	311	94.0	20	6.0
Physical pressure	High (opposition player within 2 m)	222	46.3	258	53.8
	Medium (opposition player within 5 m)	410	66.9	203	33.1
	Low (set kick/no opposition within 5 m)	836	82.9	173	17.1
Kicking foot	Preferred	1400	72.0	545	28.0
	Non-preferred	68	43.3	89	56.7
Totals		**7340**	**69.8**	**3009**	**30.2**

9780170465540

1 a What does the data show about the overall kicking effectiveness of the players in the six games?

b What types of kicks were most effective?

c What types of kicks were least effective?

d If you were a coach, what types of kicks would you focus on in practice? Why?

e If you were a coach, what types of strategies would you encourage your players to use in a game to increase their kicking effectiveness?

2 Imagine you are a performance analyst. Analyse a game or sport of your choice (live or on TV) and provide a feedback report suitable for coaching staff and/or players about the performance in the game. Pick a movement skill or tactical skill to focus on in the analysis. For example, depending on the game or sport you choose, you could analyse one of the following:

- the team's set-play performance (corners, frees, penalties)
- a team's passing performance (effective and ineffective passes)
- a player's forced and unforced errors
- the shots used by a player.

Compose your report with either a pen and paper or a software program or app.

An example of kicking performance codes is displayed in the table on page 182, which could help you develop your own analysis.

Investigation skills

WORKSHEET 8.7 IMPROVING YOUR SKILLS

Once you have analysed your performance of movement skills and tactical skills, you can adapt and refine them using practise and feedback.

Practice is the most important factor in improving your skills. Two factors that influence the effectiveness of practice are distribution and variability.

DISTRIBUTION OF PRACTICE

SB Pages 332–5

Distribution of practice is how you schedule practice.

- **Distributed practice** is practice that is more spread out.
- **Massed practice** is practice that is more bunched together.

1 For each movement skill in the table, decide whether massed or distributed practice would help you improve the skills more effectively. The first one has been done for you.

Movement skill	Massed practice	Distributed practice
Hockey goal-shooting	✓	
Sprinting		
Freestyle swimming		
Soccer passing		
Rugby goal-kicking		
Cycling		

explain to provide extra information that demonstrates understanding of reasoning and/or application

2 For each of the movement skills (including hockey goal-shooting), **explain** why you made your decision about whether massed or distributed practice would be beneficial.

____

____

____

____

____

____

____

____

____

 9780170465540

DISTRIBUTION OF PRACTICE LABORATORY ACTIVITY

Purpose

To explore the effects of massed and distributed practice on learning a movement skill.

Method

Your class will practise a movement skill using massed and distributed practice and compare them, to determine which practice schedule led to better performance.

Procedure

Half the class should practise using a massed schedule and half should practise with a distributed schedule. Complete the activity in small groups to collect the data (e.g. pairs or groups of four). Every student in the group can complete the activity in turn while other students record their scores. Collate the class scores to get a class average for each practice schedule.

Example skill

Paddle ball hit to target (this activity could be completed with another movement skill).

Equipment

Paddles (bat tennis bats), tennis balls, wall targets (e.g. concentric-circles target such as archery targets stuck to wall), stopwatches, cones (to mark hitting distance), tape measures (to measure hitting distance).

Movement skill description

- Participant is to hit (strike) a tennis ball with their non-preferred hand with a paddle so that it will hit a target as near to the bullseye as possible.
- The ball is to be bounced off the floor by the participant prior to striking it with the paddle.
- The participant is to complete 30 trials.
- The distance from the target should be around 5 m.
- The centre of the target should be approximately 1 m above the floor.

Practice conditions

- Massed practice: 5-second rest between each hit (so one member of the group has 30 tries, then a second member has all of their tries etc.).
- Distributed practice: 30-second rest between each hit (so members of the group each have one try, then each have a second try, etc.).

Results

Record your score on each hit so you can get an overall score (e.g. if using the archery target, it would be a score of 10 for the bullseye through to 1 for the outermost circle and 0 for missing the target completely). Add your score to the class data and compare the average scores for massed and distributed practice.

Score for each of 30 trials

1	2	3	4	5	6	7	8	9	10
11	12	13	14	15	16	17	18	19	20
21	22	23	24	25	26	27	28	29	30

Calculate your average score: add all your scores for the 30 trials and divide the sum by 30.

Average score =

Calculate the **average class score for massed practice:** add all of the averages for those doing massed practice and divide by the number of people doing massed practice.

Average class score =

Calculate the **average class score for distributed practice:** add all of the averages for those doing distributed practice and divide by the number of people doing distributed practice.

Average class score =

Discussion

1 Analyse the data. Which practice schedule led to better performance? Why do you think that was?

9780170465540

WORKSHEET 8.7 CONTINUED

2 Which practice schedule did people in the class prefer to practise with?

3 When do you think it would be better to use a massed practice schedule and when would be it better to use a distributed practice schedule?

VARIABILITY OF PRACTICE

Practice variability is how much skills change when you practise.

- **Blocked practice** is repetitively practising a skill for a period of time.
- **Random practice** is practising skills so that you do not practise the same skill repetitively.

3 When would you use each of these to learn a skill?

a Blocked practice

b Random practice

Investigation skills

WORKSHEET 8.8 FEEDBACK

Feedback is information you receive about your performance that can help you adapt and refine your skills to improve performance.

TYPES OF FEEDBACK

There are two main types of feedback that you can use to improve your performance:

- **knowledge of results** (KR), which is information about the outcome of movement
- **knowledge of performance** (KP), which is information about the process of skill performance that led to the outcome, such as your movement technique.

1 Classify each of the feedback in the table as either KR or KP. The first one has been done for you.

	KR	KP
A coach saying to a player, 'Keep your head up.'		✓
A player seeing the ball fly off the bat and over the fence in cricket.		
A coach saying to a player, 'That kick went more than 45 metres.'		
A physical education teacher saying to a student, 'Rotate your shoulder more.'		
A player feeling their arm move during a golf swing.		
A physical education teacher saying to a student, 'That shot was spot on target'.		

FREQUENCY OF FEEDBACK

How often you are given feedback can influence your skill acquisition. For example, your coach could give you feedback every time you hit the ball in tennis, or every fifth time, or only when you make a mistake. Giving too much feedback hinders learning, as does not giving enough.

FREQUENCY OF FEEDBACK LABORATORY ACTIVITY

Purpose

To explore the effects of different frequencies of KR feedback on performing a movement skill.

Method

Your class will practise a movement skill using different feedback frequencies and compare them, to determine which feedback frequency led to better performance.

Procedure

Practise a movement skill when intrinsic feedback is removed (using blindfolds) and when different frequencies of KR feedback are provided. The class should be divided into four different feedback frequency conditions: No KR, 25% KR, 50% KR and 100% KR. (Each student should only complete one condition.)

Complete the activity in small groups to collect the data (e.g. pairs or groups of four), with one person acting as the participant and other group members as the experimenters. Once you complete all practice trials and the transfer test, the group members can swap roles so that you all complete the exercise as a participant.

Collate the class scores to get a class average for each feedback frequency.

Example skill

Bean bag throwing (you could also complete this activity with another movement skill, such as a sofcrosse throw to a wall target, a Frisbee throw to a hoop target, or a bocce throw to a floor target).

Equipment

Blindfolds, bean bags, floor targets (e.g. archery targets stuck to floor), cones (to mark throwing distance), tape measures (to measure throwing distance).

Movement skill description

- Participant throws a bean bag so that it will hit a target as near to the bullseye as possible.
- The bean bag should be thrown underarm by the participant.
- Blindfold the participant while throwing so that they cannot see the target or the outcome of their throws.
- 50 throws to be completed.
- The distance to the target should be around 7 m.

Practice conditions

- No KR: Participants are given no feedback on the outcome of their throws.
- 25% KR: Participants are given feedback on the outcome of their throw on every fourth throw.
- 50% KR: Participants are given feedback on the outcome of their throw on every second throw.
- 100% KR: Participants are given feedback on the outcome of their throw on every throw.

When you provide KR feedback about the outcome of the throw, the participants should be told the score (e.g. 10 for a bullseye down to 1 for the outermost circle or 0 for missing the target) and also told whether the bean bag was thrown long or short or to the left or right of the bullseye.

Results

Record your score on each throw so you can get an overall score. Add your score to the class data and compare the average scores for the four feedback conditions.

Score for each of 50 trials

1	2	3	4	5	6	7	8	9	10
11	12	13	14	15	16	17	18	19	20
21	22	23	24	25	26	27	28	29	30
31	32	33	34	35	36	37	38	39	40
41	42	43	44	45	46	47	48	49	50

Your average score is the sum of all your scores for the 50 trials, divided by 50.

Average score =

The average class score for each particular KR is the sum of all averages for those doing that particular KR feedback divided by number of people doing that KR feedback.

	Average class score
No KR	
25% KR	
50% KR	
100% KR	

Discussion

1 Analyse the class data to determine which feedback frequency led to better performance. Why would this be the case?

2 Which feedback frequency did people in the class prefer?

3 When do you think it would be better to give a lot of feedback and when do you think it would be better to give less?

Investigation skills

SB Pages 340–1

WORKSHEET 8.9 TRANSFER OF LEARNING

Transfer of learning is the influence of your previous experiences on learning or performing a new skill. This means that you can transfer and adapt movement and tactical skills and strategies from one game or sport to another. For example, catching in cricket may transfer to baseball, while moving into space to get the ball in soccer may transfer to football or hockey.

1 Main tactics

Pick a game or sport you know well. Identify the main tactics in your chosen sport that might transfer to other sports.

a What sport have you chosen? What are two of its main tactics?

____

____

b What are two other games or sports that have similar tactics?

____

____

c How are those sports different to your game or sport?

____

____

2 Transfer of movement skills

a Practise a series of movement skills from one type of game (e.g. invasion games or target games) and see if any of the skills transfer and adapt to a different game. For example, you could choose net/wall games and have some members of your class practise hitting a bat tennis forehand, some practise a tennis forehand, and some practise a down ball strike.

b After a period of time practising one skill, move on to practise the other skills.

c Did any of the practice on one skill help you in performing the other skills?

____

____

d Which skill seemed to transfer the most to the other skills?

____

e Why do you think that was?

____

f Report back to the class on your findings.

Investigation skills

WORKSHEET 8.10 MODIFYING GAMES

Pages 341–5

You can modify games to challenge yourself and other participants. Aspects of the game that you can modify include the rules, number of players, space and equipment.

1 Identify whether the game modifications in the table would make the game simpler or more complex. The first one has been done for you.

	Simpler	More complex
Larger goals in soccer	✓	
Larger racquet in tennis		
Lower basketball hoop		
More players in a rugby game		
A smaller hockey field (with the same number of players)		
A rule that the team must pass the ball five times before they can shoot in basketball		
Hitting off a tee (T-ball) rather than a pitch in softball		
Using a round ball to play Australian Rules Football		
A shorter cricket pitch		

2 Choose a sport that you know well.

___

a **Design** a modified version of the sport that is simpler and suitable for children aged 10–12 years. Provide information on the rules, the number of players, and the space and equipment required.

design to produce a plan, simulation, model or similar; plan, form or conceive in the mind

___

___

___

___

___

___

___

___

___

___

WORKSHEET 8.10 CONTINUED

explain to provide extra information that demonstrates understanding of reasoning and/or application

b **Explain** why you made the modifications.

____

____

____

____

____

____

____

Investigation skills

GAME MODIFICATION PRACTICAL ACTIVITY

Purpose

To explore how modifying a game can change the way it is played.

Method

For this laboratory activity your class will participate in a game and then modify the rules of the game to see what effect this has on the game.

Procedure

Play a game (our example is tag ball) then make a rule modification and play the game again to see the effect of the modification.

Equipment

Bibs, netballs, cones, tennis balls

Game description

- Teams of four
- Playing area 15 m × 15 m (marked with cones)
- One team has the ball – they aim to pass the ball to one another so that they can tag opposition team members with the ball (not throw it at them)
- Use a netball
- Players can't run with the ball (netball rules)
- No contact is allowed
- One point is given for each successful tag
- A person cannot be tagged twice in a row

- Each team has the ball for two minutes and tries to score as many points as possible
- Then swap so the other team has the ball for two minutes
- Once each team has had a turn, make a modification to the game

Modifications to the game

- Make the playing area larger (e.g. 20 m × 20 m)
- Make the playing area smaller (e.g. 10 m × 10 m)
- Opposition players can intercept the ball (which takes one point off the score)
- Play with a tennis ball
- Players can run with the ball
- **Propose** a rule change of your own that you think would make the game harder for the team with the ball
- Propose a rule change of your own that would make the game harder for the team without the ball

propose to put forward (e.g. a point of view, idea, argument, suggestion) for consideration or action

Discussion

1 How did increasing or decreasing the size of the playing area influence the game?

____

____

2 How did allowing players to run with the ball change the game?

____

____

3 What effect did a smaller ball have on the game?

____

____

4 How did allowing the opposition to intercept the ball change the tactics?

____

____

5 What rule changes of your own did you make? How did they change the game?

____

____

____

____

____

WORKSHEET 8.11 PLAYING GAMES AND SPORTS

Pages 346–51

Playing games and sports involves competition and interaction with others. Participating requires you to:

- abide by certain game requirements so that everyone can play
- collaborate with others in participating
- motivate yourself and others
- provide leadership to others.

1 Code of behaviour

With a partner, develop a code of behaviour for participation in a specific sport. The code should include:

a how players should behave

b five principles that players should abide by in that sport.

Code of behaviour	
Sport	
How players should behave	
Example principle:	*Be a good sport.*
Principle 1	
Principle 2	
Principle 3	
Principle 4	
Principle 5	

9780170465540

2 Modifying your self-talk

Self-talk is what you say and think to yourself while performing. This can influence your motivation and confidence.

a In the table below, list some common negative thoughts you have when playing games and sports. One example has been given.

b For each negative thought, write down a positive statement you could use to replace it. For example, 'I can't do it' can be replaced with 'Most players have difficulty with this, but I have done it before'.

Negative self-talk	Positive self-talk replacement
What if I mess up the shot?	You have taken this shot successfully lots of times before.

WORKSHEET 8.11 CONTINUED

3 Sportspersonship

Provide three examples of how you can demonstrate good sportspersonship in a game. One example has been given.

	Description
Example	*Shaking hands with your opponent after the game*
Sportspersonship example 1	
Sportspersonship example 2	
Sportspersonship example 3	

4 Sportspersonship research

Interview a classmate, friend or someone you know about an incident in which someone displayed bad sportspersonship in a game.

a What was the incident?

____

____

____

b What did the interviewee do?

____

____

____

____

c What did other people at the game do?

____

____

____

____

d How did it make the person feel?

e What could be done to reduce the chances of such an incident happening again?

5 Leadership: umpire a game

Leadership in games and sports can involve taking on different roles as part of the game. One of these roles is as an umpire/referee/official.

a As part of a class have each student take a turn as umpire of a game.

b Reflect on the umpiring. Did you enjoy performing the role? What did you find difficult in the role?

CHAPTER 8 REVIEW

Place a tick next to the skills you have learnt while completing this chapter of the workbook.

Now I can:

- ◯ Identify the essential elements of games and sports.
- ◯ Compare different types of games and sports.
- ◯ Examine differences between fundamental and specialised movement skills, gross and fine movement skills, and open and closed movements skills.
- ◯ Investgate the use of tactical skills in games and sports, including tactics, strategies and game plans.
- ◯ Analyse, adapt and refine movement skills and tactical skills in games and sports.
- ◯ Analyse, adapt and refine practice and feedback to enhance your performance and the performance of others in games and sports.
- ◯ Transfer and adapt skills and strategies from previous experience to ne movement situations.
- ◯ Modify games to vary complexity, including modifying the rules, the number of players, and the space and equipment.
- ◯ Reflect on how ethical behaviour, fair play, codes of behaviour, sportspersonship, teamwork, motivation and confidence, leadership and collaboration skills can have an impact on participation in games and sports.

9780170465540

9

PHYSICAL ACTIVITY PLANS

Shutterstock.com/Nicetoseeya

WORKSHEET 9.1 DAILY PHYSICAL ACTIVITY

Pages 356–9

There are five ways you can incorporate movement opportunities when planning your personal physical activity plan. You can add these opportunities by:

- maximising incidental physical activity (e.g. walking around a shopping centre, walking between classes, walking from the carpark to the school classroom)
- doing household chores and yard work (e.g. vacuuming, raking leaves, etc)
- increasing occupational physical activity (e.g. activity during school time)
- active commuting (e.g. walking, skating or riding to school)
- increasing leisure-time physical activity (includes structured sport or exercise for fitness).

1 Insert each of the five movement opportunities in the diagram below. The first one has been provided as an example.

2 Near each box, sketch a picture depicting an example activity for each movement opportunity.

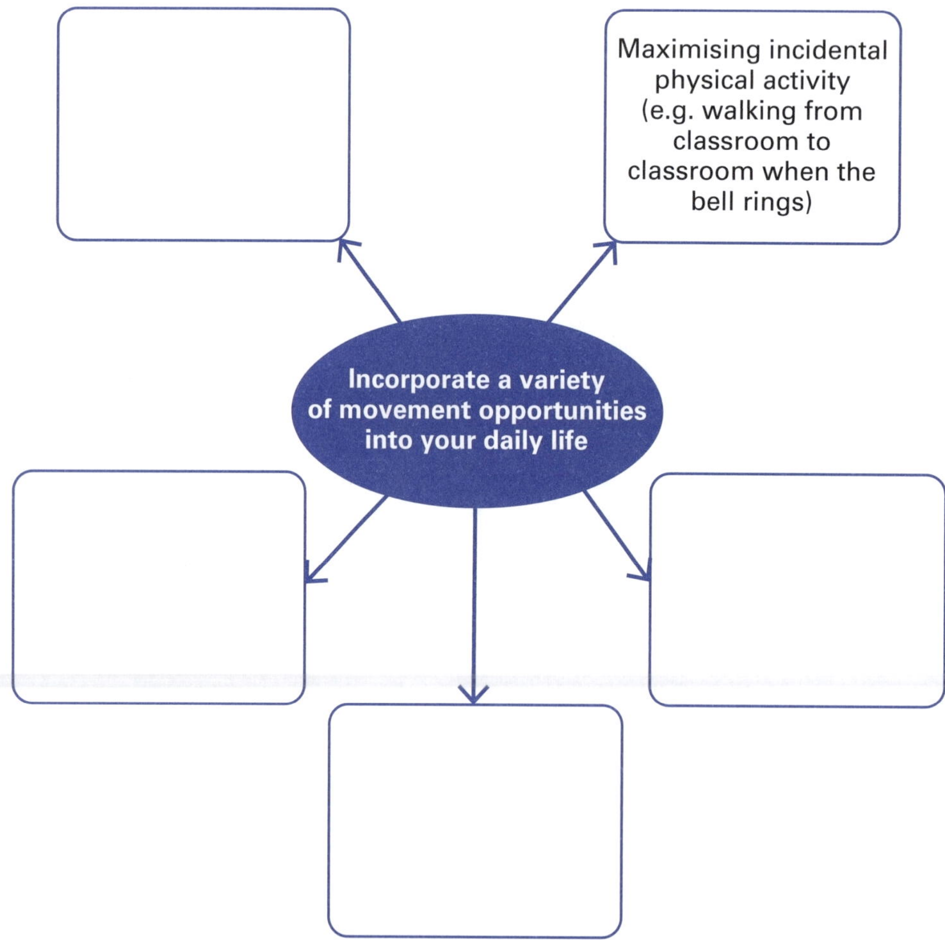

WORKSHEET 9.2 NEEDS ANALYSIS

Page 358

Complete the following needs analysis to help you to develop your personalised physical activity plan. The table below will assist you to complete your needs analysis.

	Strength training	Power training	Muscular endurance training
% of 1 RM*	80–95	30–50	40–60
Sets	3–5	3–5	2–4
Reps	2–4	3–5	15–20
Speed	slow/medium	fast	medium/fast
Rest	3–4 mins	3–4 mins	1–2 mins

*1 RM = the maximum amount of weight a person can lift in a single repetition for a given exercise

1 Set your goals (e.g. to have more energy to do things after school or to have more energy during the last half of a soccer match)

____

____

2 Current physical activity level

Tick which of the following relates to you:

☐ Last week I *did not* meet the Australian 24-hour Movement Guidelines for Children and Young People (5–17 years).

☐ Last week I *did* meet Australian 24-hour Movement Guidelines for Children and Young People (5–17 years).

You can also tally your daily pedometer steps

__________ = Day 1 pedometer steps in total

__________ = Day 2 pedometer steps in total

__________ = Day 3 pedometer steps in total

__________ = Day 4 pedometer steps in total

__________ = Day 5 pedometer steps in total

__________ = Day 6 pedometer steps in total

__________ = Day 7 pedometer steps in total

...	+	...	+	...	+	...	+	...	+	...	+	...	=	...
Day 1		Day 2		Day 3		Day 4		Day 5		Day 6		Day 7		Total

...	÷	7	=	...
Total number of steps				Average daily pedometer step count

3 Establish your priorities

Rank your priorities (e.g. 1 = highest priority; 5 = lowest priority) by completing the table below.

Rank	Physical outcomes	Rank	Social and psychological outcomes
	Aerobic capacity		Feeling positive
	Flexibility		Feeling happy
	Muscular strength		Being active with friends
	Muscular endurance		Being active with family
	Body composition		Feeling more confident
	Muscular power		
	Balance		
	Speed		
	Agility		
	Reaction time		
	Having more energy		

4 Fitness testing

Based on your priorities, identify five fitness tests you could complete to assess your physical fitness. Next to the health-related components, place the letter (H) and an (S) next to the skill-related components.

a ______________________________

b ______________________________

c ______________________________

d ______________________________

e ______________________________

9780170465540

5 Tailor your priorities to your interests

Insert your five main interests (e.g. spending time with my friends, training for football) in the spaces provided below.

a ______________________________

b ______________________________

c ______________________________

d ______________________________

e ______________________________

6 Availability of time and resources

a Use colour coding in the table of commitments below to denote when you have the following commitments. For example:

- red = family commitments
- yellow = school classes (e.g. English, maths, HPE, etc)
- blue = other commitments
- pink = sporting commitments
- green = time available to be active.

Commitments and opportunities to be active							
	Monday	Tuesday	Wednesday	Thursday	Friday	Saturday	Sunday
Before school							
During school							
After school							
Evening							

b List equipment you have access to at home to assist you to be active (e.g. exercise bike, skipping rope, weights, bicycle, roller skates, hula hoops).

____

____

____

____

7 Appropriate training methods

Use the information table below to identify the training methods you could use to develop your five top-priority fitness components. Then complete the table on the next page.

Fitness component training method	Aerobic power	Local muscular endurance	Muscular strength	Anaerobic power	Muscular power	Agility	Flexibility	Speed
Energy system	**Aerobic energy**		**ATP-PC energy***					**Anaerobic glycolysis****
Continuous	✓	✓						
Fartlek	✓	✓						
High Intensity Interval Training (HIIT)	✓	✓						
Long interval	✓	✓						
Intermediate interval								✓
Short interval				✓				✓
Speed				✓				✓
Weights/ resistance			✓	✓	✓			
Circuit		✓	✓	✓	✓	✓		✓
Plyometrics				✓	✓	✓		✓
Flexibility							✓	
Swiss ball and Pilates		✓	✓				✓	

* ATP-PC energy system provides the most rapid source of energy and does not rely on oxygen. The more intense the activity the more ATP-PC used (however, exercise usually only lasts for about 20 seconds at maximum intensity).

** Anaerobic glycolysis – sometimes referred to as the lactic acid system – makes energy without oxygen. It can make twice as much energy than the ATP-PC system but at a slower rate.

9780170465540

Complete the table below using your student book or research.

Priority areas	Description of priority (e.g. fitness component)	Suitable training method
Example	*Speed*	*Short interval training*
1		
2		
3		
4		
5		

8 Physical condition of the performer

To determine if you are ready to begin a training program, complete a Physical Activity Readiness Questionnaire (PARQ). Search online for 'Par-Q and You' to find an online version of the questionnaire.

a Are you injury free? (circle answer)

Yes/No

b Are you physically ready to start a new physical activity program? (circle answer)

Yes/No

By completing this needs analysis you have determined:

- your goals
- your current physical activity level
- your priorities to focus on
- the fitness tests that you could complete to determine a baseline level of fitness
- your interests
- the resources and time you have available to be active
- appropriate training methods that could be used to focus on your priority areas
- your readiness to commence a personal physical activity and fitness program.

This can be used to determine improvements achieved in subsequent assessments.

Equipped with the information from your needs analysis and the 10 principles of planning training programs you learnt about in your student book (specificity, intensity, duration, frequency, progressive overload, detraining, maintenance, individuality, diminishing returns and variety), as well as with the help of your teacher, you are now ready to start designing your own personal program.

WORKSHEET 9.3 FITNESS TEST ANALYSIS

Pages 358–81

How do you measure improvement or determine areas you need to work on?

To evaluate how effective your personal activity plan is, you need to assess whether your fitness or physical activity level has increased, decreased or remained the same, compared with before you started the new program.

To establish this starting level (baseline) of fitness, you need to complete some fitness tests. The fitness tests are then repeated and your results pre-test, during the program and post-test (at least 12 weeks after commencing your program) or at the end of a program are analysed.

Although the ideal would be to stay on a fitness program for the rest of your life, the reality is that most people struggle to stick to a new fitness program, even for 12 weeks. So, to last three months would be an excellent outcome.

Research fitness tests used to assess various health and skill-related components of fitness.

1 Complete several (at least five) fitness tests (your teacher may determine the tests that need to be completed).

2 Complete the following for the fitness tests you have done.

  a Test name: ______________________

  Fitness component assessed: ______________________

  Result: ______________________

  Rating compared to norms for your age (if available): ______________________

  Area requires (circle one): Maintenance (2 sessions/week)/Improvement (3+ sessions/week)

  b Test name: ______________________

  Fitness component assessed: ______________________

  Result: ______________________

  Rating compared to norms for your age (if available): ______________________

  Area requires (circle one): Maintenance/Improvement

  c Test name: ______________________

  Fitness component assessed: ______________________

  Result: ______________________

  Rating compared to norms for your age (if available): ______________________

  Area requires (circle one): Maintenance/Improvement

d Test name: ____________________

Fitness component assessed: ____________________

Result: ____________________

Rating compared to norms for your age (if available): ____________________

Area requires (circle one): Maintenance/Improvement

e Test name: ____________________

Fitness component assessed: ____________________

Result: ____________________

Rating compared to norms for your age (if available): ____________________

Area requires (circle one): Maintenance/Improvement

f Test name: ____________________

Fitness component assessed: ____________________

Result: ____________________

Rating compared to norms for your age (if available): ____________________

Area requires (circle one): Maintenance/Improvement

g Test name: ____________________

Fitness component assessed: ____________________

Result: ____________________

Rating compared to norms for your age (if available): ____________________

Area requires (circle one): Maintenance/Improvement

WORKSHEET 9.4 TRAINING PRINCIPLE DEBATE

Pages 301–6

1 In small groups, investigate one of the 10 key training principles – specificity, intensity, duration, frequency, progressive overload, detraining, maintenance, individuality, diminishing returns and variety.

2 In pairs or small groups you will prepare a mini debate with another group. The topic is:

(Insert assigned training principle) is the most important training principle in fitness improvement.

Each team will have a set amount of time to prepare and present their argument. You could also allow time for a rebuttal. Ensure you include the key fitness components that could be developed in your argument.

3 Play a game of charades in a small group. Take turns acting out an assigned training principle.

WORKSHEET 9.5 FITNESS AND WELLBEING PROGRAMS IN YOUR COMMUNITY

1 On the example timetable below, place a tick next to each type of activity you have previously participated in.

Group fitness timetable

Start time	Monday	Tuesday	Wednesday	Thursday	Friday	Saturday	Sunday
6.00 a.m.	Functional training (Gym)	Cycle	Yoga	Pilates	Body pump		
8.00 a.m.	Senior Strength Express	Zumba Gold	Seniors Strength	Boxing/ Circuit		Les Mills CX WORX	
9.00 a.m.	Les Mills CX WORX	Les Mills CX WORX			Functional training (Gym)	Les Mills Body Pump	Les Mills Body Pump
9.30 a.m.	Aqua-Lo	Body Blitz	Les Mills CX WORX	Les Mills Body Step	Les Mills Body Pump	Aqua	Les Mills RPM
10.30 a.m.	Les Mills Body balance	Yoga	Boxing	Les Mills RPM	Pilates	Les Mills Body Attack	Les Mills Body Step
11.30 a.m.	Les Mills RPM	Tai Chi	Yin Yoga	Aqua Gentle	Tai Chi	Yoga	Cycle
4.30 p.m.		Teen fit	Teen fit	Meditation	Zumba		Women's only Yoga
6.00 p.m.	Aqua Tone	Functional training (Gym)	Les Mills CX WORX		Les Mills RPM	Aqua Tone	Pilates
6.30 p.m.	Les Mills Body combat	Les Mills Body Attack	Les Mills Body Pump	Boxing	Les Mills Body Pump		
7.30 p.m.	Les Mills Body Pump	Zumba	Les Mills Body Step	Pilates	Yoga		
8.30 p.m.	Yoga	Pilates	Les Mills Body balance	Yin Yoga	Deep Water Aqua		

2 Read the descriptions of the classes in the following two tables and then research your local community fitness and recreation centre to determine which are offered. Use the last column in the tables to indicate whether each program is offered or not.

Freestyle classes	Description	Offered at my local community fitness centre (circle)
Aqua	Improve your fitness and endurance with this dynamic and exciting pool-based class for all fitness levels. (45 mins)	Yes / No
Aqua-Lo	Light to moderate-intensity pool-based workout. (45 mins)	Yes / No
Aqua Tone	A total toning aqua class incorporating kickboards noodles and aqua dumbbell. (45 mins)	Yes / No
Body Blitz	A high-intensity cardio and circuit workout targeting your total body and core strength. (55 mins)	Yes / No
Boxing/ Circuit	This energetic class is made up of different boxing and circuit style exercises designed to help reduce weight loss and improve muscle tone and increased fitness. (55 mins)	Yes / No
Deep Water Aqua	Aqua class in deep water using floatation belts with a focus on toning and core strength. (55 mins)	Yes / No
Functional Training	Incorporates TRX, kettlebells, slam balls and more to improve basic functional movements. (45 mins)	Yes / No
Gentle Aqua	Low impact water workout- great introduction to aquatic exercise or for those returning to exercise or for rehabilitation as it is gentle on the joints. (45 mins)	Yes / No
Meditation	A class that incorporates a state of deep peace when the mind is calm and relaxed. (25 mins)	Yes / No
Pilates	These sessions work on the body's core strength and stability to promote correct posture and alignment. (45 mins)	Yes / No
Seniors Strength	A resistance training workout using a variety of styles designed specifically for mature adults only. (45 mins)	Yes / No
Tai Chi	A Chinese system of slow meditative physical exercise designed for relaxation, balance and health. (55 mins)	Yes / No
Teen Strength	Learn safe techniques to move safely as you build your strength and fitness. Incorporating bodyweight and light resistance to build strong foundations for weight training and sports. (45 mins)	Yes / No
Yin Yoga	A slow paced style of yoga with postures or asanas that are held for longer periods. (55 mins)	Yes / No
Yoga	Help to increase flexibility, strength and joint stability. Reduces physical and mental stress. (55 mins)	Yes / No
Zumba®	The Zumba® program fuses hypnotic Latin rhythms and easy to-follow dance moves. (55 mins)	Yes / No

Les Mills classes	Description	Offered at my local community fitness centre (circle)
Body Attack®	A high-energy, calorie burning athletic workout with strong, simple moves to pumping music. (55 mins)	Yes / No
Body Balance®	Uses a range of movements and motion set to music that will improve your mind and body. Blends elements of Yoga, Pilates & Tai Chi. (55 mins)	Yes / No
Body Combat®	Combines moves and stances developed from self-defence disciplines – Karate, Boxing & Muay Thai. Fiercely energetic class taught in a safe, simple manner aimed at burning fat. (55 mins)	Yes / No
Body Pump®	A rapid, resistance training workout using barbells providing the quickest fat-burning fitness class. (55 mins)	Yes / No
Body Step®	High energy cardio combination targeting abdominal area, buttocks and thighs. (55 mins)	Yes / No
CX Worx®	A high-intensity workout that will tighten and tone core muscles, improves functional mobility and strength. (25 mins)	Yes / No
RPM/Cycle	Indoor studio cycle class. Aims to burn 600 calories in 45 mins. / Low impact workout with the intensity controlled by you. (45 mins)	Yes / No

3 **Discuss** which age groups the programs target within the community based on the offerings shown.

____

____

____

discuss
to talk or write about a topic, taking into account different issues or ideas

4 Evaluate the advantages and limitations of a class like 'Body Step' for:

a yourself

b an elderly person

____

____

____

WORKSHEET 9.5 CONTINUED

describe
to give an account of characteristics or features

AC

5 Analyse the timing of when certain programs are offered and **describe** why you think certain activities are offered at certain times of the day. Consider the target groups in your discussion.

6 Explain why Gentle Aqua would be suitable for older adults.

7 Select two programs that you think would be of interest to you and specify two major training principles that are likely being used and three fitness components likely to be developed. An example is shown in the table below.

Program	Training method 1	Training method 2	Fitness component 1	Fitness component 2	Fitness component 3
Boxing circuit	*Circuit*	*High-Intensity Interval Training (HIIT)*	*Muscular power*	*Muscular endurance*	*Aerobic capacity*

9780170465540

WORKSHEET 9.6 FARTLEK TRAINING

Page 367

Shutterstock.com/Spaceport9

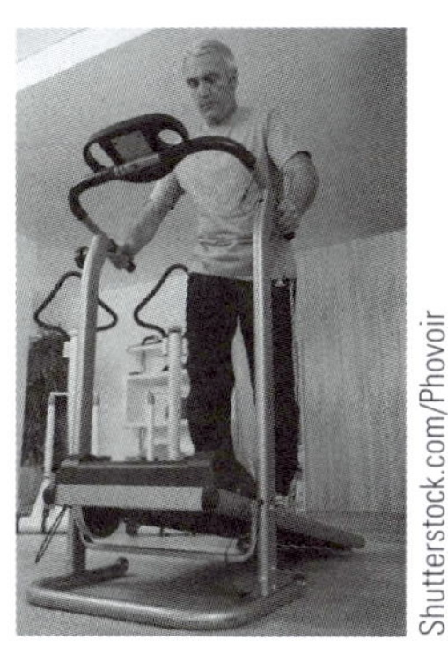
Shutterstock.com/Phovoir

Treadmills use Fartlek training programs by simulating the inclines of hilly terrain.

The diagram below represents undulating terrain where a person could run or cycle, either out in the natural environment in the bush, or the man-made built environment around the streets. Exercise equipment such as treadmills, exercise bikes and stair climbers with onboard computers can simulate the terrain by making it easier or harder for you as you exercise.

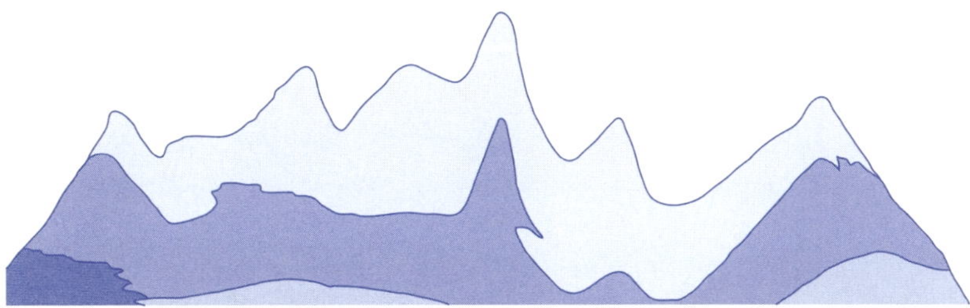

1 Plot the relevant intensity using lines on the cross-section of terrain on the diagram above. Use a key to denote the likely intensity required to cover over the terrain:

__________ = low intensity (e.g. walk or slow jog)

__________ = high intensity, short bursts (e.g. sprint or going uphill).

Keep in mind that it is very dangerous to sprint downhill, not only because of the increased risk of tripping and falling because of your faster speed, but also the forces on your joints (especially the knee) can be eight times greater than on level ground.

2 **Describe** what is meant by the term interval training.

____

____

describe
to give an account of characteristics or features

AC

3 Describe either a built environment or natural environment near your home or school where you could complete Fartlek training either on foot or on a bike.

____

____

____

WORKSHEET 9.7 STRETCHING ACTIVITY

Page 370

1 Participate in a range of static stretching activities either on your own or in class. You may need to research some ideas for different stretches online. Which stretches did you use?

____

____

____

2 On the figure below, label and colour the muscle groups that were being used in each stretch. Use different colours for different stretches.

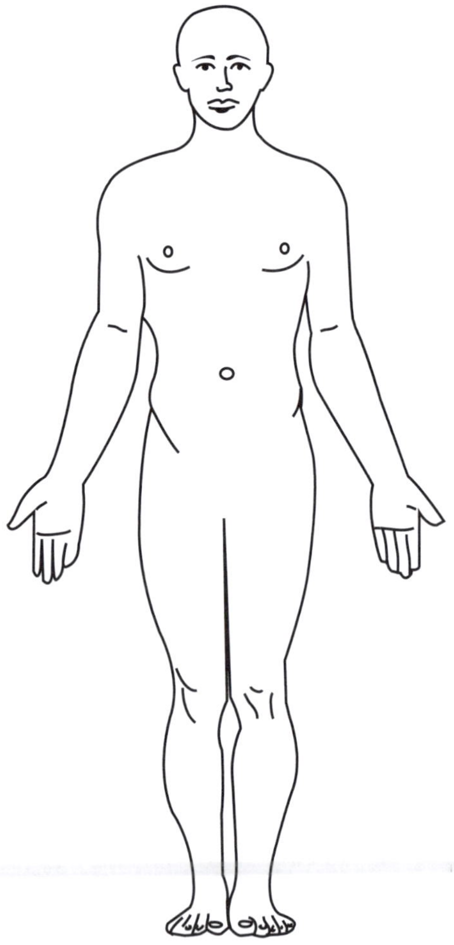

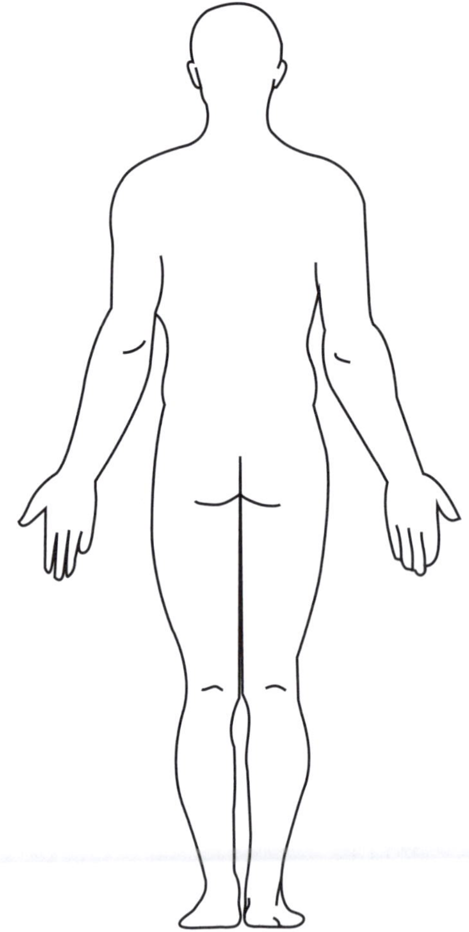

9780170465540

3 Define each of the following:

a **Dynamic stretching**

b **Static stretching**

c **Proprioceptive neuromuscular facilitation (PNF) stretching**

4 Research two PNF stretches for your legs and **describe** how to perform these two stretches.

describe
to give an account of characteristics or features

AC

5 **Explain** how being flexible reduces your risk of injury.

explain
to provide extra information that demonstrates understanding of reasoning and/or application

AC

6 Does having good flexibility in your hamstrings guarantee good flexibility in your quadriceps? Explain your answer.

WORKSHEET 9.8 FITNESS FUN

Pages 370–90

explain
to provide extra information that demonstrates understanding of reasoning and/or application

1 Search online for a cartoon relating to a fitness training principle or a training method.

2 **Explain** how it relates to either a training method or a training principle. Use the cartoon below as an example.

'Actually, I'd rather chase the ball than go for a walk. Sprinting is the best way to firm up flabby glutes!'

3 Design your own cartoon depicting a training principle or training method.

WORKSHEET 9.9 WEEKLY TRAINING SCHEDULE

Page 373

Complete the weekly training schedule for yourself in the table below. Insert all physical activities you complete in a typical week, including:

- active commuting (e.g. walking or cycling to school)
- household chores (e.g. garden work, vacuuming, hanging out washing)
- leisure time activities (e.g. walking the dog, netball training, going to the gym)
- school-based physical activity (e.g. PE classes, sport class, soccer training).

If you currently only do physical activity while you are at school, try to design an ideal weekly training schedule as a starting point for the next two weeks. Make sure it is realistic and something you could achieve and build upon.

	6–9 a.m. (before school)	9 a.m.–3.30 p.m. (during school)	3.30–10 p.m. (after school)
Monday			
Tuesday			
Wednesday			
Thursday			
Friday			
Saturday			
Sunday			

1 How many minutes per week do you spend on the following:

a Active commuting: ______________________________

b Household chores: ______________________________

c Structured organised sport (e.g. sports training or classes): ______________

d PE/sport at school: ______________________________

WORKSHEET 9.10 FITNESS EQUIPMENT

Pages 376–8

Review the examples of three types of non-specialised fitness and play equipment.

Household items	Recycled items	Inexpensive sporting items
Blanket	Tyre	Frisbee
Tarp	Box	Tennis ball
Broom	Milk crate	Pool noodle
Broom handle	Chaff sack	Skipping rope
Bucket	Rope	Play ball
Small steps	Milk or cordial bottle	Hacky sack
Hay bale	Sand	Hula hoop

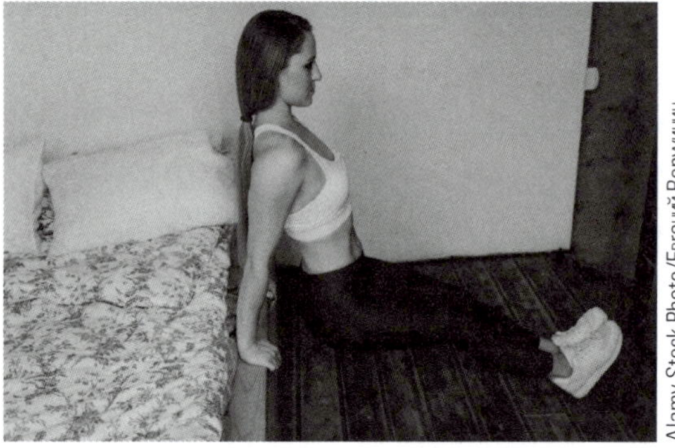

Alamy Stock Photo/Евгений Верминин

iStock.com/LightFieldStudios

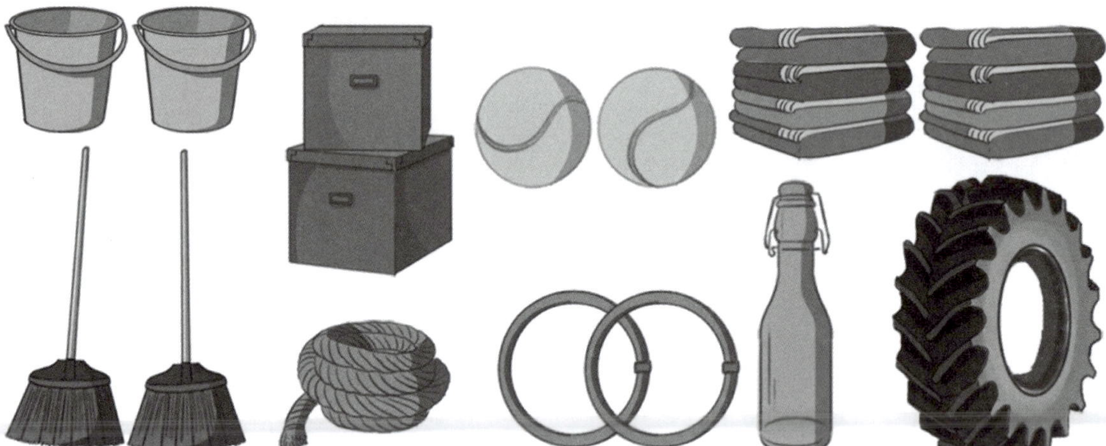

You don't need specialised or expensive exercise equipment to be active at home.

DESIGN YOUR CIRCUIT

1 Using at least three items from each of the three columns in the table, design a fitness circuit consisting of eight stations. You can use the template on the next page to help you.

- The circuit stations need to consist of an activity you could complete for 60–90 seconds before rotating to the next station.
- You should complete the circuit two or three times.
- Think about the order of your activities. Ensure you do not include two stations in a row that require the same muscle groups be used. Alternate stations to allow certain body parts to rest while others can work hard.

PARTICIPATE IN YOUR CIRCUIT

- Make sure you try your circuit out. You could do it at home while listening to music.

2 Draw your equipment and describe the exercise/activity to be performed at each station in each box. Assume you will move in a clockwise direction.

1	2	3
8		4
7	6	5

EVALUATE YOUR CIRCUIT

describe
to give an account of characteristics or features

a **Describe** how well the circuit worked.

______

______

______

______

discuss
to talk or write about a topic, taking into account different issues or ideas

b **Discuss** whether you were tired after doing the circuit.

______

c Did you enjoy the circuit?

______

d Did you do the circuit on your own or with someone else? (If so, who?)

______

______

explain
to provide extra information that demonstrates understanding of reasoning and/or application

e Outline two changes **explaining** what you could modify in your design to make the circuit even more effective and enjoyable.

______

______

______

______

______

______

9780170465540

WORKSHEET 9.11 WHAT HELPS MOTIVATE YOU?

9

SB
Pages 378– 81

Describe three strategies you could use to motivate yourself to work hard when you are being active.

describe
to give an account of characteristics or features

AC

1 ______________________________

____

____

____

____

2 ______________________________

____

____

____

____

3 ______________________________

____

____

____

WORKSHEET 9.12 TRADITIONAL INDIGENOUS GAMES RESOURCE

1 Search for a game for students in a similar year level to you using the filter and select a game category (e.g. target games).

Game selected:

____

Weblink
Check out the Yulunga Traditional Indigenous Games resource

2 Select a game you would like to participate in and outline the background, language, a short description, players needed, playing area, duration, equipment and game play and basic rules (e.g. 'kungirruna' p131 of the Yulunga Traditional Indigenous Games resource).

____

____

____

____

____

____

____

3 Participate in the game you selected. Describe what you enjoyed about learning a new physical activity.

____

____

4 Describe the social benefits associated with participation in this activity selected.

____

____

____

5 Using the Yulunga resource identify five other traditional Indigenous games you would like to try.

____

____

____

____

____

CHAPTER 9 REVIEW

Place a tick next to the skills you have learnt while completing this chapter.
Now I can:

- ☐ Participate in a range of physical activity options to design, implement and evaluate participation strategies that promote health and social outcomes.
- ☐ Incorporate lifelong physical activities and structured physical activities into my daily life.
- ☐ Complete a needs analysis when planning a personal activity program.
- ☐ Analyse fitness test data.
- ☐ Develop a weekly training schedule.
- ☐ Describe the 10 key training principles of specificity, intensity, duration, frequency, progressive overload, detraining, maintenance, individuality, diminishing returns and variety.
- ☐ Use various training methods such as Fartlek and flexibility training and develop circuits using non-specialised equipment.
- ☐ Participate in physical activities designed by First Nations Peoples.

10

FIT TO DANCE

Shutterstock.com/Nicetoseeya

9780170465540

WORKSHEET 10.1 DANCE AND SPORT

SB Pages 395–6

1 How do you define fitness?

2 Explain the principle of 'specificity' in fitness training.

3 List what you think are the core components of fitness required by elite-level dancers, stating why each component is necessary. Provide an example where possible.

Elite-level dancers		
Fitness components required	**Explanation of why this fitness component is necessary**	**Example, where possible**

WORKSHEET 10.1 CONTINUED

4 List the core fitness components required by an elite 100-metre track sprinter, stating why each component is necessary. Provide an example where possible.

Elite athlete		
Fitness components required	**Explanation of why this fitness component is necessary**	**Example, where possible**

5 Analyse the similarities and differences between these two elite level athletes.

6 Class discussion

Share your ideas with the rest of your class.

WORKSHEET 10.2 INTERVIEW WITH BEN DAVIS

Ben Davis, a former soloist with the Australian Ballet, danced at an elite level for more than 10 years. He then went on to work as a trainer at the Ballet Theatre Australia.

Page 396

Newspix/Katrina Tepper

Ben Davis and Olivia Bell rehearse *Dyad 1929*, watched by choreographer Damien Welch at Australian Ballet headquarters in Melbourne.

Fairfax Syndication/Simon Schluter

Ben Davis and Amy Harris rehearsing *Cinderella* for the Australian Ballet.

Read the transcript below of an interview with Ben Davis when he was professionally dancing. Then answer the questions that follow.

INTERVIEW

Lee Anton-Hem **(LA-H):** How long have you been dancing and how much of that time has been at an elite level?

Ben Davis **(BD):** I started dancing around the age of seven, first just as a hobby, and then one class a week turned into two, became three, and so on, until I was practising every night of the week and all day Saturdays. When I was 16, mum and dad allowed me to go to an institution that combined both my academic and ballet training and the rest is history. I danced overseas briefly from the age of 18, came home, and this is my ninth year with the Australian Ballet. All up, it's been about 24 years that I've been dancing and 12 of those years have been at an elite level.

(LA-H): So you've been used to working hard for a very long time?

(BD): It's very much part of the everyday 'grind'!

(LA-H): Can you tell us about the physical demands of a ballet performance?

(BD): They change nightly, depending on the ballet we're performing and what roles you are performing, but pretty much every muscle is worked in ballet. It doesn't matter whether you are doing a large role or a small role – it's pretty physical and you can be doing it for a three-hour period each night.

(LA-H): When the Australian Ballet is performing, how many performances would you give in a week?

(BD): Usually seven performances in a week, but if there's something that is popular, something that people really want to see, we can do up to eight performances a week.

(LA-H): And that doesn't include your training time, or your rehearsal time, does it?

(BD): No, it doesn't. Training usually starts at 11 in the morning; class runs for an hour and 15 minutes, then goes into a rehearsal period through to the afternoon. We have a short dinner break and then people might go to the gym or do some cycling, then it's back to the theatre around five or six o'clock to put stage makeup on, another half-hour warm up class, then into the performance, which can go sometimes till 10 or 10.30 at night. So we're looking at 11- to 12-hour days and we're probably active for 6 to 8 hours of that day.

(LA-H): Wow! So let me get this right, you're physically active for around 6 to 7 hours – that's basically a full day's work for most people – that's a little bit more than the average person like me would need to do to stay fit and healthy – just a little bit!

(LA-H): We know that dancers, and particularly ballet dancers, are exceptionally fit athletes. Which of these areas [cardiorespiratory endurance, anaerobic power, muscular strength, flexibility, body composition] would ballet dancers require high levels of fitness in, to be able to dance at an elite level?

(BD): Probably all of the areas – ballet tends to be quite anaerobic; you are doing a lot of fast movements or you need a lot of energy for short periods of time, but then again, you need to be able to do those over a long period of time, depending on the length of the performance. Our training is usually anaerobic: we are doing a lot of small exercises, one after the other, but you have a little rest in between. But there are some ballets that will require you to be on stage and active for 20 or 30 minutes, non-stop.

(LA-H): So you require the whole gamut of fitness – all the components, across the board?

(BD): Yes!

(LA-H): In terms of your cardiorespiratory endurance, what sort of training do you undertake to be able to perform at this level?

(BD): Our ballet training of a morning does aid with that, but a lot of dancers find that that's not quite enough, so cycling's great training, or swimming; a lot of people just like to be at the gym.

(LA-H): How do you train for your anaerobic power?

(BD): The main one would be, if we are on the bike, to mix up the speeds: keeping a steady pace and then every few minutes or so really going hard for a minute – really pushing our bodies to get the heart rate back up, and then bring it back down to a steady level.

(LA-H): Do you find that helps, when you're on stage, and you might be in the middle of a performance and all of a sudden you have a solo part, where you've got to do leaps and jumps – does that help prepare for that?

(BD): Definitely – there are times on stage when you might be quite static for a period of time, then you have to get up and have to do these fantastic things, so it's definitely aiding in that.

(LA-H): But I suppose nothing would beat actually doing that in your warm-up or rehearsals?

(BD): Nothing quite matches our daily class – it really provides us with everything we need.

(LA-H): What other training do you do?

(BD): I like to do gym work, a good combination of upper body and lower body, usually leg press and also lifting weights, to help with lifting the girls – if you can do it at the gym, then lifting them will be a breeze!

(LA-H): When you are dancing in a pas de deux (dancing with a partner), you have to do static lifts and ballistic lifts. Static lifts are when your partner is standing by you and you have to lift her up. The ballistic lifts are when your partner might be running towards you and you have to catch her and lift her up. How do you prepare for that?

(BD): It's just practice. Pas de deux work is usually quite an even amount of work between the male and the female. She is holding her centre to help aid the lift, but a ballistic lift is usually a little easier, as she is using her legs and her feet and can actually jump to help get that lift up into the air.

(LA-H): Do you find that weight training and pushing the weights in the gym actually does help?

(BD): Definitely – if you can do it in the gym, then doing it with someone who is holding themself, and is not a deadweight, is much easier.

(LA-H): How about flexibility training? How do you get your legs so high in arabesques and standing splits?

(BD): A lot of ballet training is about lengthening our limbs and making the lines quite long, so it's not about gripping our muscles to lift our legs. Everything we do is about lengthening our spine, lengthening our muscles, so it just becomes second nature to do these kinds of movements after a while.

(LA-H): How long do you need to train for to get yourself into a standing split position? I wouldn't imagine that it's something that you can just do over a couple of days?

(BD): For some people, it's very, very easy; they would be able to do it without a problem at all. Others, like myself, have to work really hard to get those long lines.

(LA-H): Things like stretching and Pilates must form a huge part of your training regime, because flexibility is so important to a dancer. Can you give us an idea of how much time you would spend in a day just working on your flexibility?

(BD): A lot of dancers will arrive before class even starts to either do a stretching warm up or a Pilates session, which is all about lengthening the muscles, not so much about gripping and stretching in a static position for too long, but keeping the lengthening through a movement. So you might have your leg up on a barré and rather than just pushing down and holding it, we're stretching over the leg and then coming back up. We're constantly trying to get length through the muscles, without pushing to their extreme too much.

(LA-H): As a performer, it's really important for you to listen to your body and train sensibly, so that you don't get injured. So part of you being able to train, and being able to back it up each day with a performance, would involve your recovery. What do you do to recover?

(BD): Hydration is really important, as well as replenishing our muscles, so making sure we've got a good sensible, healthy diet to back that up, and also getting as much rest as possible. Sleep is when our body repairs. So if you can put all those things together, it really aids in getting you ready for the next day as well.

(LA-H): There's so much that goes into it, isn't there? It's not just about the training, the dancing, it's about preparing your body in other ways as well.

(BD): Exactly. Some of the things we can do are ice baths at the end of each rehearsal and at the end of each performance, and a lot of dancers like to wear compression tights or stockings between rehearsals to help the muscles recover that little bit quicker.

(LA-H): Ben, what would you eat, in a normal day?

(BD): I try to have a large breakfast in the morning, to get my body started. It's usually a bowl of cereal of some sort, or something with bran in it. Depending on rehearsal schedules, sometimes there isn't time –you might have 10 minutes here, 15 minutes there, so anything small that will give you that boost of energy you need.

(LA-H): And the slow-release carbohydrates?

(BD): Definitely: fruits, nuts, that sort of thing. Sometimes you just need a quick sugar fix as well, to give you a burst of energy – jellybeans are great!

(LA-H): How do you feel about having a career in dance?

(BD): Extremely lucky. I get to work with very talented people at the top of their field each day, and also it's allowed me to travel the world while I'm doing what I love. Some people only ever dream of doing something like this, so I do feel extremely lucky.

(LA-H): So it's very rewarding and fulfilling?

(BD): Definitely, very rewarding

(LA-H): Well we all can't be amazing dancers like you, but what about dance as a hobby or means of keeping fit? What would you say to the average student out there who may never have participated in dance? Would you encourage them to give it a try?

(BD): Absolutely! We've all danced at some point, even if it's secretly in front of our bedroom mirror! And dancing is fun, whether you are doing it as an elite athlete or just as a hobby.

(LA-H): And it's a great way to improve your fitness, too.

(BD): Definitely – it gets the heart rate going and you get the happy endorphins as well.

(LA-H): Did you know that some football and rugby clubs teach dance to improve coordination?

(BD): I had heard this, yes. I like that. I bet they like it, deep down, too!

(LA-H): When I think about the really great benefits you can get from dance, I worry still that boys don't take up the opportunity as much. Do you think that's because of old stereotypes or do you think that's changing now?

(BD): It's changing, with shows like *So You Think You Can Dance*, and *Dancing with the Stars*. People are becoming a little more educated about the types of dance that are out there and seeing how athletic it is. And that it's about having fun and being fit.

(LA-H): Genres such as hip hop as well have made dance more appealing to the average person.

(BD): That's the great thing about dance, that's there's so many different genres, so there's something for everyone's taste.

(LA-H): Well, Ben I would love to talk to you all day, but we better let these students get back to their work or we will be in trouble with their teachers!

1 How much time does Ben spend in training each day?

2 Which particular components of fitness did Ben say that ballet dancers require high levels of to be able to dance at an elite level? Which components are more prominent than others?

3 Other than dance class and performances, what sort of training does Ben do to work on his cardiorespiratory endurance?

4 Other than dance class and performances, what does Ben use to work on his anaerobic power?

5 As well as practising lifts with his partner in class, how does Ben enhance his muscular strength?

WORKSHEET 10.2 CONTINUED

6 What does *pas de deux* mean?

7 In dance, what is the difference between a static and a ballistic lift?

8 Flexibility is a major fitness component for ballet dancers. What sort of flexibility training does Ben do?

9 What does Ben do to assist his body's recovery? Include the types of foods Ben eats as well.

10 Do you think there are still stereotypes around dancing or do you think that things are changing? We heard Ben's response. What is yours?

WORKSHEET 10.3 HEALTH-RELATED FITNESS

Page 398

As a class, perform *The Nutbush*. Record your resting heart rate in the space below before you begin and then again after you have completed the dance. Chat with a partner during the dance if you wish; dancing should be fun!

Resting heart rate before *The Nutbush* = ____________________

Heart rate after performing *The Nutbush* = ____________________

Answer the following questions after you perform the dance and have recorded your heart rate.

1 Was your heart rate higher after you completed *The Nutbush*? If so, why?

____

____

2 Were you able to have a conversation with your partner while you were dancing?

____

3 Explain what happened to your body during the dance.

____

____

____

____

4 What exercise intensity level did your body begin to work at?

____

5 What intensity level do you think the majority of people would work at when performing *The Nutbush*?

____

6 Did you enjoy doing this type of physical activity? Why/why not?

____

____

____

____

____

9780170465540

WORKSHEET 10.4 COUNTING

Page 400

1 a With your partner, perform the first eight counts of *The Nutbush* to establish the first eight-count movement sequence.

b Once you have done this for the first eight counts, continue doing the dance for the second, third and fourth group of eight counts.

c Document each movement sequence below.

Groups of eight counts	Description of eight-count movement sequence
1	
2	
3	
4	

2 Why do you think the movement sequences are based on eight counts?

___

___

3 Does the dance provide the same amount of movement on the right and left side of the body?

___

4 Do you think the amount of movement on both sides of the body is important when exercising? Why/why not?

___

___

5 What is different about the last group of eight counts?

___

___

9780170465540

WORKSHEET 10.5 CREATING A NEW MILD

SB Pages 401–5

1 Working cooperatively with a partner, think, pair and share your ideas to come up with four new movement sequences to replace the existing movement sequences in the original dance.

When creating your new moderate intensity line dance (MILD), it needs to:

- consist of four movement sequences with equal number of counts. *The Nutbush* uses eight counts for each movement sequence but you can use 16 or 32 counts as long as the counts are the same for each movement.
- consist of movement on the right and left side of the body and opposing muscle groups
- continue the moderate intensity level of the dance.

What I think	What my partner thinks	What we will share

2 **Extension activity**

When doing *The Nutbush*, the dancer turns anticlockwise to face each wall. You can choose to add this element to your MILD or keep the dance facing the front.

WORKSHEET 10.6 COMPLETING YOUR MILD

Page 404

1 Now that you and your partner have developed the footwork for your moderate intensity line dance (MILD), you can extend it by adding in specific arm movements for each of your moves.

2 Think of a name for your MILD and record your dance in the following table.

Title:		
Groups of eight counts	**Description of movement sequence**	**Arm movements**
1		
2		
3		
4		

3 With your partner, teach your MILD to two other students and then learn their dance to reflect on the differences between both routines.

WORKSHEET 10.7 PROVIDING FEEDBACK

Page 405

Dancing and how we express ourselves through dance is very personal and unique to each individual. Therefore, when we work in groups, it is important to remember that we each have our own dance style and that we should not judge each other harshly. However, we can provide feedback to our peers about whether they meet the set criteria for a given task.

1 When reflecting on a MILD, your reflection should only be about the following criteria.

   Did the MILD contain four new movement sequences:
   - consisting of equal counts?
   - consisting of movement on the right and left side of the body?
   - continuing the moderate intensity level of the dance?

2 Complete the reflection below for the MILD you learnt from the other two students.

MILD criteria	Yes or no
Did the MILD consist of four new moves?	
Did each move consist of equal counts?	
Did each move consist of movement on the right and left sides?	
Did the MILD continue in moderate intensity?	

WORKSHEET 10.8 TAI CHI

Pages 408–9

Read the two articles and then answer the questions that follow.

Tai Chi and millennials

Though studies show that most of those who practice tai chi are 50+ years old, many instructors report a renewed interest among younger people looking for a stress-release option. Tai chi master Terry Dunn, who has been teaching tai chi in Los Angeles for more than 30 years, says that in the past 18 months he has received significant increased enquiries, especially from young men in the IT industry. Google headquarters has been offering tai chi to its employees for the past couple of years. Master David Chang, owner of the Wushu Central Martial Arts Academy in San Jose, confirms he has several students in their 20s and 30s. Taniela Irizarry, 37, says it helps with her injuries from years of beating up her body as a professional dancer.

Most health research into the benefits of tai chi focus on people over 50. But, like many physical activities, the earlier you start in life, the more you reap the rewards, says Dr. Michael Irwin, director of UCLA's Mindful Awareness Research Center. Benefits are cumulative, and if stress can be dealt with in a positive way at a younger age, it will not accumulate over time and become much more serious in terms of detriments to health.

Irwin has carried out more than a dozen peer-reviewed studies evaluating the ability of tai chi, yoga and mindfulness to improve health outcomes. Irwin's earliest tai chi study in 2003 found an increase in the number of disease-fighting 'T cells' that fight off shingles in patients who practised tai chi. The 15-week study showed no increase in those who did not participate. In a later study, Irwin found that practising tai chi gave the same immunity boost as a vaccine developed for shingles. In 2015, a study found that tai chi, over several weeks, reduced cellular inflammatory responses in patients suffering from insomnia. Irwin is now studying whether tai chi can be as effective as cognitive behavioral therapy to treat depression and anxiety.

Source: 'Tai chi fights stress, getting popular with Millennials', Amy Chillag, 5 September 2017, CNN Health

Study highlights the benefits of Tai Chi and yoga in aged care

Implementing modified tai chi and yoga programs into aged care facilities enhances residents' quality of life through improved physical, emotional and intellectual wellbeing, according to Auckland University of Technology research.

The research published recently in the Journal of Clinical Nursing aimed to identify the appropriateness and acceptability of modified tai chi and yoga in Australian aged care facilities.

The program was designed by tai chi and yoga trainers, researchers in senior health and general nursing and aged care staff to ensure its safety and that it suited frail residents and their individual abilities.

The research, conducted in a 108-bed aged care facility in Newcastle, NSW, involved 16 residents aged 66 to 92, three aged care staff and half-hour tai chi and yoga classes twice a week for 14 weeks.

Lead researcher Dr Padmapriya Saravanakumar said residents found the program and activities appropriate, feasible and holistic for the mind and body.

'This is significant especially as older adults in residential aged care settings have fewer active opportunities owning to age and health-related functional limitations,' Dr Saravanakumar told Australian Ageing Agenda.

The overall theme of comments from participants at the completion of study was an appreciation of the mind-body approach of the exercise, the study found.

In feedback, residents involved the study described the programs as slow and mindful, new and exciting, gentle but rewarding and calming and relaxing, said Dr Saravanakumar, a lecturer in nursing at the school of clinical sciences at Auckland University of Technology.

Prior to the research, the aged care facility had offered other exercises to residents including seated group physical activities and access to a gym, however most felt they were only able to participate to some extent due to fears of falling and exhaustion, she said.

'The unique mind-body approach of yoga and tai chi and the sense of safety it provided must have helped the participants view this holistic exercise positively,' Dr Saravanakumar said.

She said the study found tai chi and yoga benefited residents through improved confidence, balance, posture and body awareness.

'The participants reported experiencing benefits from the exercise and enjoyment from the company of other residents. They also felt a sense of purpose in engaging actively, contributing suggestions to the instructors and in looking out for their peers,' Dr Saravanakumar said.

She said supporting research and evidence-based interventions are among the ways aged care facilities can explore opportunities to provide active ageing opportunities for residents.

'Study highlights benefits of tai chi and yoga in aged care' by Sandy Cheu, *Australian Ageing Agenda*, 12 September 2018

1 Read the articles above, then use your own research skills to find two more articles that highlight the benefits of Tai Chi. Attempt to find at least one article that has been written by a medical practitioner or allied health professional, or that has been based on research findings. Write the details of the articles and their authors below and analyse the findings.

2 Create a brochure to promote Tai Chi as a form of rhythmic movement for improving joint mobility and other health-related concerns. These brochures can be handed out at your dance for fitness session. When creating your pamphlet, remember to acknowledge your sources of information (if relevant).

Front

Back

9780170465540

WORKSHEET 10.9 MINDFUL STRETCH ROUTINE

SB Pages 406–7

Your teacher will allocate you and your partner a muscle group to design a static stretch for. Use these stretch diagrams as a starting point.

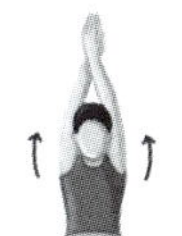
Latissimus dorsi and posterior deltoid stretch (link hands, push elbows together)

Triceps stretch (pull elbow across and down)

Shoulder rotator stretch (using towel, pull up with the top arm then down with the other)

Pectoral stretch at 90° and 120° (use doorway or post)

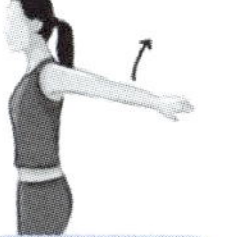
Biceps stretch (hands apart)

Supraspinatus stretch (keep elbow parallel to ground)

Wrist extensor stretch (tilt head to opposite side, keep elbow straight)

Thoracic extension stretch (reach forward with arms, push chest towards floor, arch back down, bottom behind knees)

Lateral flexion stretch (one side, then the other, push pelvis across as you bend)

Lumbar flexion stretch (be gentle if sore)

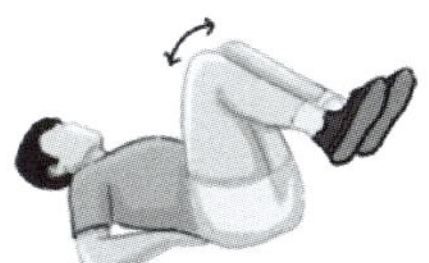
Lumbar rotation stretch (rotate legs one side, then the other side, draw in and brace stomach muscles at the same time, breathe)

Hamstring stretch (straighten leg)
i. with foot pointed
ii. with foot pulled back towards the knee

Hamstring stretch (commence with knee slightly bent, then push knee straight as tension allows, push chest toward foot)

Adductor stretch (push down with elbows on knees very gently, keep back straight)

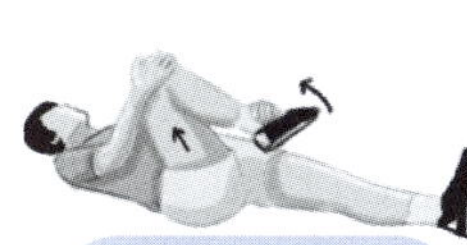
Gluteal stretch (pull knee and lower leg towards opposite shoulder)

Gluteal and lumbar rotation stretch

Quadriceps stretch (keep pelvis on floor)

Quadriceps stretch

Adductor stretch (keep foot pointing forward, lunge sideways on bent knee, keep back straight)

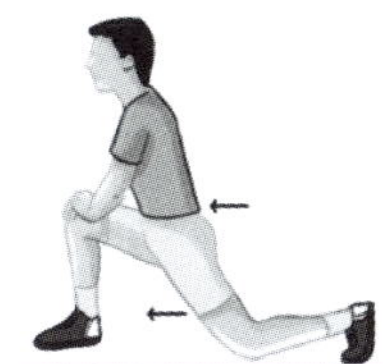
Hip flexor stretch (keep back straight, tuck bottom under, lunge forward on front leg)

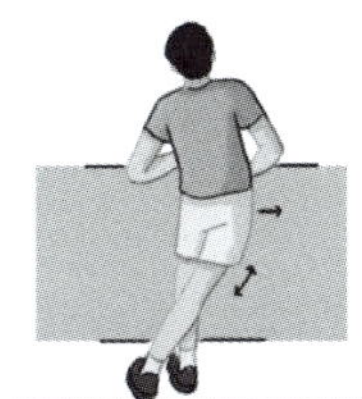
Tensor fascia stretch (continue to push bottom forward, while pushing hip to the side)

Gastrocnemius stretch (keep knee straight and heel down, feet facing forward)

1 With your partner, create a slow, controlled and coordinated static stretch for each opposing muscle in the group. You will also develop a script around the delivery of your stretch for the rest of your class about the mindful performance of the movement involved.

For example, a quadriceps stretch would be followed by a hamstring stretch.

Each stretch will be:

- static and must not bounce
- performed in a slow and controlled manner
- lengthened to the point of resistance, not pain

- held for 10 to 20 seconds
- repeated three times
- followed by slow rhythmical movement to proceed into the opposing muscle stretch
- performed on a stable, safe and shock-absorbing surface (not concrete!).

Remember that safe technique is very important and a stretch should never be forced. It is important to perform stretching on a safe surface. You need to have a mat to protect your back and knees from the ground.

2 Research your stretch with your partner and document the stretch in the table below. You can draw diagrams to assist your explanation. You can use this stretch also in your dance for fitness session.

3 Research how to teach a mindful stretch. Things to consider in your mindful stretch script include:

- Description of how to safely perform the stretch
- Use of the breath – inhaling and exhaling
- Kinesthetic sensation through your senses, such as feeling, seeing, hearing
- Relaxation of the muscle group

Write your mindful script in the table below.

Stretch name:	
Muscle or muscle group	**Opposing muscle or muscle group**
Script for how to perform the stretch.	**Script for how to perform the stretch.**

4 Teach another pair your stretch using your mindful approach.

9780170465540

WORKSHEET 10.10 DANCE FOR FITNESS

SB Pages 413–6

Together as a class and with your teacher, plan a class dance for fitness session. You will need to decide when and where the session will occur and the format of the event. You can then develop flyers and posters to advertise your session.

DANCE FOR FITNESS SESSION PROMOTION

1 As a class, discuss and reflect on the health benefits of dance for people of your own age group and together decide which of these benefits would be appropriate to promote as a part of your dance for fitness session.

2 Note down the following information:

a the main health benefit(s)

b information your participants will need to be able to attend your session, such as:

- time
- location
- what to wear
- what to bring; such as water bottle and asthma puffers if necessary.

WORKSHEET 10.10 CONTINUED

3 Using the spaces provided below, design a poster or flyer to advertise your session. Then create the poster/flyer, make copies and display them in the most appropriate places around the school to generate maximum attendance.

Front

Back

4 With your teacher's guidance, think about other opportunities to promote your dance for fitness session within the school. Make a note of these opportunities below.

5 Conduct your dance for fitness session. Have fun!

DANCE FOR FITNESS SESSION EVALUATION

Reflect on and evaluate the effectiveness of your dance for fitness session in the space below. Consider the following points.

- Did you provide a moderate intensity fitness workout for participants?
- Did you provide a safe environment and activity for participants?
- Did you promote physical activity through dance?
- Did you provide a physical activity opportunity through dance?
- Did you provide a safe stretch routine at the end of the session?
- Did the participants have fun?

CHAPTER 10 REVIEW

Place a tick next to the skills you have learnt while completing this chapter of the workbook.

Now I can:

- ◯ Identify and explain the fitness components within dance.
- ◯ Compare/contrast the fitness requirements of an elite-level dancer to other elite athletes.
- ◯ Explain how dance can assist in developing individual and community health-related fitness.
- ◯ Use existing knowledge of fitness principles to redesign an existing social dance into a moderate intensity training dance.
- ◯ Develop an aerobics-style workout that could be used in my own personal physical activity program or for a before- or after-school program to promote physical activity opportunities for all students.
- ◯ Understand and implement safe dance practices.
- ◯ Examine the knowledge and skills of partners and group members when redesigning and performing an existing social dance to increase the cardiovascular output from moderate intensity to moderate/vigorous intensity level.
- ◯ Use decision-making strategies to work cooperatively and constructively with others when making decisions about movements, tempo, music and placement.
- ◯ Evaluate dance routines based on set criteria.
- ◯ Investigate the social and cultural practices of physical activities, such as Tai Chi.
- ◯ Create a safe and slow stretching-based routine to improve flexibility

Now reflect upon your learning in this chapter by completing the following sentences.

1 Information I found interesting was …

____

____

____

____

2 I was surprised to learn ...

3 Information I found useful was ...

4 I developed strength in ...

5 I am looking forward to ...

6 I could change my lifestyle by ...

7 This could benefit others by ...